Development and Politics from Below

Non-Governmental Public Action

Series Editor: **Jude Howell**, Professor and Director of the Centre for Civil Society, London School of Economics and Political Science, UK

Non-governmental public action (NGPA) by and for disadvantaged and marginalized people has become increasingly significant over the past two decades. This new book series is designed to make a fresh and original contribution to the understanding of NGPA. It presents the findings of innovative and policy-relevant research carried out by established and new scholars working in collaboration with researchers across the world. The series is international in scope and includes both theoretical and empirical work.

The series marks a departure from previous studies in this area in at least two important respects. First, it goes beyond a singular focus on developmental NGOs or the voluntary sector to include a range of non-governmental public actors such as advocacy networks, campaigns and coalitions, trades unions, peace groups, rights-based groups, cooperatives and social movements. Second, the series is innovative in stimulating a new approach to international comparative research that promotes comparison of the so-called developing world with the so-called developed world, thereby querying the conceptual utility and relevance of categories such as North and South.

Titles include:

Barbara Bompani and Maria Frahm-Arp (*editors*)
DEVELOPMENT AND POLITICS FROM BELOW
Exploring Religious Spaces in the African State

Jude Howell and Jeremy Lind
COUNTER-TERRORISM, AID AND CIVIL SOCIETY
Before and After the War on Terror

Jenny Pearce (*editor*)
PARTICIPATION AND DEMOCRACY IN THE TWENTY-FIRST CENTURY

Non-Governmental Public Action Series
**Series Standing Order ISBN 978–0–230–22939–6 (hardback) and
978–0–230–22940–2 (paperback)**
(*outside North America only*)

You can receive future titles in this series as they are published by placing a standing order. Please contact your bookseller or, in case of difficulty, write to us at the address below with your name and address, the title of the series and the ISBN quoted above.

Customer Services Department, Macmillan Distribution Ltd, Houndmills, Basingstoke, Hampshire RG21 6XS, England

Development and Politics from Below

Exploring Religious Spaces in the African State

Edited by

Barbara Bompani
Teaching Fellow, Centre of African Studies, The University of Edinburgh, UK

Maria Frahm-Arp
Senior Lecturer, School of Theology, St Augustine College, South Africa

First published 2010 by
PALGRAVE MACMILLAN

Palgrave Macmillan in the UK is an imprint of Macmillan Publishers Limited,
registered in England, company number 785998, of Houndmills, Basingstoke,
Hampshire RG21 6XS.

Palgrave Macmillan in the US is a division of St Martin's Press LLC,
175 Fifth Avenue, New York, NY 10010.

Palgrave Macmillan is the global academic imprint of the above companies
and has companies and representatives throughout the world.

Palgrave® and Macmillan® are registered trademarks in the United States,
the United Kingdom, Europe and other countries.

ISBN-13: 978–0–230–23775–9 hardback

This book is printed on paper suitable for recycling and made from fully
managed and sustained forest sources. Logging, pulping and manufacturing
processes are expected to conform to the environmental regulations of the
country of origin.

A catalogue record for this book is available from the British Library.

A catalog record for this book is available from the Library of Congress.

10   9   8   7   6   5   4   3   2   1
19   18   17   16   15   14   13   12   11   10

Printed and bound in Great Britain by
CPI Antony Rowe, Chippenham and Eastbourne

Jacket photo: Prayers at Diamalaye Mausoleum in Dakar, Senegal 2006.
Photo by Emilie Venables.

Contents

Foreword

The average volume of conference proceedings is like the bag of miscellaneous items often found at auctions: a collection varied in origin, purpose and utility. The common factor is that all are of a size to fit the bag; the attraction to the bidder usually in the desirability of particular items rather than in the contents as a whole. Not so with this volume. Variety there certainly is: far-reaching theoretical expositions and painstaking detailed case studies; variety of viewpoint, variety of discipline, variety of locality (for the articles reflect the life of societies right across Africa).

But there is also coherence and direction; the various objects in the bag are not miscellaneous, they belong together. They form a set of tools for multiple use.

One major sphere of use, as the title suggests, is in the troubled area of development studies and development practice. Here the volume makes a major contribution by demonstrating the vast range of indigenous forces already producing societal change. One of the contributors quotes a conversation heard between the members of a Muslim women's group in Mali: 'this is what our group is about, this is what our attempts to practice humanity are for. We want to change the current state of social ills.' In this case, as in so many others in the book, the forces for change are religious. This is also a book about religion, and its place in the public sphere; a book about religious innovation and religious conservation, about old traditions and movements of renewal. It makes clear the ineffectual nature of Western models of development that assume that religion belongs properly to the private sphere, and should be kept there. The uncomfortable truth is that the 'modern' Western model of the universe is simply too small for Africa. For practical purposes, it has bracketed out of consideration huge areas of experience that are a vital part of the consciousness of the greater part of humanity. Meanwhile, the student of religion will gain from the book further evidence of the extent to which Africa is becoming a major theatre of dynamic religious thought and activity, vital to the future of both Christianity and Islam. Certainly there is much here to show that African Christianity can be equally viewed either as the African chapter of the history of Christianity or as the Christian chapter in the history of African religion;

while the studies from Islamic communities show similar combinations of the historic and universal with local modes of appropriation and transmission.

A grateful reader must thank editors and contributors for a rich collection. This is certainly no bag of oddments; more a set of stones that together make a fine necklace.

Andrew F. Walls
University of Edinburgh, Liverpool Hope University,
and Akrofi-Christaller Institute, Ghana

Acknowledgements

In April 2008 the Centre of African Studies at the University of Edinburgh, and the Wits Institute for Economic and Social Research (WISER) at the University of the Witwatersrand, hosted a joint conference in Edinburgh exploring religious spaces in Africa by engaging in a discussion on the relationship between development, politics and religion in the continent. Three key issues emerged. The first was that the religion, politics and development needed to be drawn together into a conversation in which it was recognised that they impacted on each other in quite profound ways. The second point was the need for research to be conducted which moves beyond the teleology of scientific, modern, individualism to allow us to study spaces from a different perspective – one in which community, the spiritual and factors within the human experiences such as love, death, loss, prosperity and power can be seen not only through a lens of causal, rational choices. And the third related issue was the new perspective on Western, secular, externally driven development that the new thinking on the relationship between religion, politics and development present. Bound up with this is the question of who offers 'development' and how religions or governments have and have not either helped or hindered 'development' programmes.

This volume brings together a unique grouping of colleagues from universities in Europe, USA and Africa. It has been a great pleasure to work with such a varied group of scholars from different disciplines and with expertise in different African countries. We are grateful for their support, hard work and patience in the production of this edited work. In addition we would like to thank our colleagues at the Centre of African Studies in Edinburgh University and at WISER, Wits University, for the generous support and encouragement in the organisation of the conference and the book project. We owe particular thanks to David Maxwell who introduced us in Johannesburg in 2007, which ultimately led to this fruitful collaboration.

This work would not have been possible without funding provided by the British Academy Conference Grant, the ASA-UK conference grant, the generous contribution of the Moray Endowment Fund, College of Medicine and Veterinary at the University of Edinburgh and the Binks

Trust of the University of Edinburgh. We are also very grateful to the Economic Social Research Council (ESRC), in particular through the ESRC Non-Governmental Public Action programme.

Finally particular gratitude must be expressed to our partners, James Smith and Raj Rajakanthan for having been extremely supportive and patient while we were working on this volume. The endless cups of tea and unfailing humour were much appreciated. This volume has been developed and realized through long skype phone calls and emails between Edinburgh and Johannesburg, and sometimes skype fails and emails do not arrive, this perhaps also speaks to the problematics of North–South relationships and their over-reliance on technology and modernity!

Barbara Bompani and Maria Frahm-Arp
Edinburgh and Johannesburg

Notes on Contributors

Barbara Bompani is Teaching Fellow at the University of Edinburgh. Her publications include 'African Independent Churches in Post-Apartheid South Africa: New Political Interpretations', *Journal of Southern African Studies* (2008); 'Mandela Mania: Mainline Christianity in Post-Apartheid South Africa', *Third World Quarterly* (2006). Her interest focuses on the production of knowledge around faith, development and the relationship between civil society, society and politics in Southern Africa.

Ezra Chitando works for the Ecumenical HIV and AIDS Initiative in Africa (EHAIA), a programme of the World Council of Churches (WCC). He is an Associate Professor in History and Phenomenology of Religion at the University of Zimbabwe. He has published widely on the Church's response to HIV and AIDS in Africa and on gender. His works include *Living with Hope: African Churches and HIV/AIDS* (2007), *Singing Culture: A Study of Gospel Music in Zimbabwe* (2002).

James R. Cochrane is Professor in the Department of Religious Studies, University of Cape Town, co-Principal of the African Religious Health Assets Programme (ARHAP) and Director of the Research Institute on Christianity and Society in Africa (RICSA). His publications include *Circles of Dignity: Community Wisdom and Theological Reflection* (1999); he is editor with Bastienne Klein of *Sameness and Difference: Problems and Potentials in South African Civil Society* (2000) and with Gary Gunderson he is writing *Religion and the Health of the Public: Conceptual Foundations*.

Stephen Ellis is Desmond Tutu professor in the Faculty of Social Sciences at the Vrije Universiteit Amsterdam and senior researcher at the African Studies Centre in Leiden, the Netherlands. With Gerrie ter Haar he wrote *Worlds of Power: Religious Thought and Political Practice in Africa* (2004). His most recent book is *Madagascar: A Short History* (2009) co-authored with Solofo Randrianja.

Maria Frahm-Arp is a senior lecturer at St Augustine College in South Africa where she teaches Church History and the Sociology of Religion. Her particular interests are in women and religion, medieval

Christianity and contemporary Pentecostal Charismatic Churches in Southern Africa. Her book *Professional Women in South African Pentecostal Charismatic Churches* is forthcoming.

Elizabeth Graveling completed her PhD, entitled 'Negotiating the Powers: Everyday Religion in Ghanaian Society', at the University of Bath in 2008. She currently lectures in the field of Development Studies within the Department of Social and Policy Sciences at the University of Bath. Her research interests fall in the areas of religion and development, and the sociology and anthropology of contemporary religion (particularly Christianity) in Africa and the United Kingdom.

Ernest T. Mallya is Associate Professor of Public Policy and Administration at the University of Dar es Salaam (UDSM) Tanzania and the Deputy Principal responsible for Academics at the Dar es Salaam University College of Education (DUCE). His publications include *'Aids, Poverty and Representative Democracy in Tanzania' Afriche e Orienti* (2009), 'The Political Economy of Democracy in Tanzania' *Journal of African Elections* (2007). His research interests include professionalism and ethics in government; political parties, democracy, good governance and the non-profit sector.

Dorothea E. Schulz is Professor of Social & Cultural Anthropology at the University of Cologne, Germany. Her publications include *Perpetuating the Politics of Praise: Jeli Praise Singers, Radios and Political Mediation in Mali* (2001); 'The world is made by talk: female youth culture, pop music consumption, and mass-mediated forms of sociality in urban Mali', *Cahiers d'Etudes Africaines* (2002). Her forthcoming book is entitled *Pathways to God: Islamic Revival, Mass-mediated Religiosity and the Moral Negotiation of Gender Relations in Urban Mali*.

David E. Skinner is Professor of History at Santa Clara University and a member of the Advisory Board for *Critical African Studies*. He published 'The Incorporation of Muslim Elites into the British Administrative Systems of Sierra Leone, The Gambia and the Gold Coast', *Journal of Muslim Minority Affairs* (2009) and with Jalloh Alusine, *Islam and Trade in Sierra Leone* (1997). For more than 40 years his research has examined the educational, economic and political roles of Islamic organizations in North and West Africa.

Abdulkader Tayob is Professor of Religious Studies at the University of Cape Town. His publications include *Religion in Modern Muslim Discourse* (2009); *Islam in South Africa: Mosques, Imams and Sermons* (1999);

and forthcoming *The Shifting Politics of Identity: In Islam and Modernity*. His interests include Islam in African public life, and modern Islamic thought.

Gerrie ter Haar is Professor of Religion and Development at the Institute of Social Studies, the Hague. She is Vice-President of the International Association for the History of Religions (IAHR) and founding member of the African Association for the Study of Religions (AASR). Her most recent book is *How God Became African: African Spirituality and Western Secular Thought* (2009). She is co-author of 'The Role of Religion in Development: Towards a New Relationship between the European Union and Africa' (2006), which won an essay prize awarded by the *European Journal of Development Research*. Her current project is an edited volume on religion and development (forthcoming).

Linda van de Kamp is a cultural anthropologist and a researcher at the VU University Amsterdam and the African Studies Centre in Leiden. Her specific fields of interest are religion, South-South transnational connections, violence, risk, gender and reproductive issues. She is currently finishing her PhD dissertation 'Big War against Maledictions': The Violence of Conversion to Brazilian Pentecostalism in post-War Urban Mozambique'.

Introduction

Development and Politics from Below: New Conceptual Interpretations

Barbara Bompani & Maria Frahm-Arp

If we gaze through the narrow teleological prisms of politics, development, progress and modernization, then religions should have disappeared by now. This perspective, that for such a long time dominated Western and some Non-Western thought, never came true. The secularization paradigm, based on European history and the modernist model of the progressive exclusion of religion from the public sphere, has seen religion as an obstacle to progress, to be sidestepped, ignored or eliminated. After decades of exclusion from political and developmental analyses, a shift has taken place and over the last 15 years we have begun to discern a continuing or, in other cases, renewed interest in the role of religion in public life.

Literature on religion as a social and political actor is emerging today from several different perspectives. Development agents and political analysts have started to seriously consider the role played by religious values and religious organizations in supporting or hindering policies and processes. A set of literature is emerging that analyses religious organizations *in* development (cf. Berger, 2003; Marshall & Keough, 2004; Tyndale, 2006; Haynes, 2007; Marshall & Van Saanen, 2007; Clarke & Jennings, 2008; Birmingham RAD Programme, http://www. rad.bham.ac.uk/index.php?section=1). Furthermore, after 9/11 religion has become a key focus in political science, and literature has been produced on the role of religion, especially Islam, in the political sphere (Berger, 1999; Juergensmeyer, 2000; Mamdani, 2004; Norris & Inglehart, 2004; Roy, 2004; Stout, 2004; Fetzer & Soper, 2005; Roy, 2007). However, and it bears underlining, many studies have primarily focused on untangling the negative potentialities of religion as a generator of conflict. Finally, over the last 20 years academics have turned their attention to the rise and transformation of religion and religiosity around the world,

1

especially in the Global South. Protestantism has seen a remarkable growth in Latin America, to the detriment of Catholicism; new forms of Charismatic Christianity have developed in this continent, as well as in Africa and Asia; Islam, in various forms, is growing in Africa and Asia and a rising number of diasporic groups, with their religious networks and views, are changing the religious milieus of Western societies (e.g. Martin, 1990; Stoll, 1990; Allen, 1992; Casanova, 1994; Haghayeghy, 1996; Peterson, 1996; Heelas, 1998; Hansen, 1999; Corten & Marshall-Fratani, 2001; Romero, 2001; Martin, 2002; Brass, 2003; Anderson & Tang, 2005; Soares & Otayek, 2007). In the first decade of the twenty-first century we can see, sense, and believe that not only has religion not disappeared, but that it has re-emerged reinvigorated and in new forms and patterns.

A striking factor regarding this recent proliferation of debates, studies and publications (and funding!) on the role of religion in development and politics is that the approaches used to investigate religion are usually analytical tools derived from a Western context and as such reflect Western interpretations of what religion does, or can do, for development and politics. This has underpinned studies that have categorized religious organizations and groups either as those who can deliver development and operate as development agencies or as religious mainline or fringes groups that are potentially dangerous and politically organized, especially in relation to Western norms and conceptualizations of modernity (cf. Alkire, 2006; Clarke & Jennings, 2008). These perspectives are invariably viewed through a Western lens. While much of this literature needs to be investigated, analysed and understood, it is clear that what is missing , and becoming an obstacle to our investigation and understanding the role of religion in development and politics, is a genuine understanding of the meaning of religion itself and the way that it is embedded in the everyday life of millions of people in the South. There has been an overwhelming focus on the consequences but not on the causes, on what religion does but not what religion is and this implies an a priori exclusion of a deeper understanding of the role of religion in contemporary societies. Ultimately, what does it mean to be religious in non-Western countries? Is it different from being religious in the Global North? Are Western developmental agencies and Western political analysts keen (and ready) to acknowledge that? What conceptual and analytical tools do we need to understand religiosity? And if the secularization model holds true in certain respects in the West but not in many other parts of the world, should we not then talk of the uniqueness of the West (Davies, 2002) and perhaps therefore acknowledge the need to produce new interpretative models that do not stem from European/North American contexts?

In this book we aim to move beyond a narrow conceptualization of the separation between public and private spheres, and beyond a limited understanding of development and politics, and in doing so offer an interpretation of how development and politics are played out in the everyday lives of religious people, who, we must not forget, represent the great majority in contemporary African societies. To work meaningfully with religion, politics and development and their interactions we have sought to engage in a conversation that incorporates philosophy, theology and an analysis of political and developmental power structures. We have also tried to understand how people – ordinary people living in urban, rural and peri-urban spaces in the South – think about themselves and their place in a modernizing, hybridizing world. We have engaged with the influences of local ideas of modernity, development and representation, ideas about the individual and new forms of communication. These different elements cannot be seen in isolation but only in relation to the community and to one another.

In this volume we have treated and used development and politics in a broad 'non-conventional' way: we have questioned formal and informal public arenas, reflected on the meaning and power of symbols and rituals within organized as well as unorganised groups, and analysed social action promoted by religious ideas and religious people. This offers a way of understanding religion and its potential in the local context and develops an understanding of societies that use the lens of religion as a fundamental component of social construction and social interpretation. Social and political changes, as well as development implementation, need to be accepted and internalized by the people concerned (Rist, 2002). To borrow Jean-François Bayart's expression, development and politics should come from 'below' (Bayart, 1993).

We are not concerned with producing a functional categorization of religious organizations as development actors or analysing the impact or 'quality' of religious organizations from a Western developmental perspective (cf. Clark and Jennings, 2008). Instead, we seek to provide a much-needed reflection on the significance of the kinds of public action that manifest themselves in non-Western societies. Religion needs to be understood in its entirety, not just from the point of view of its contribution (or opposition) to development and political establishments. Development and political studies often deny the need for a holistic analysis, and often religion and beliefs are the first variables to be removed from analyses (Deneulin & Bano, 2009). This is particularly important in contexts in which separation into spheres, for example the public and the private, has never been applied to religion and in which religion is perceived in a holistic way. Furthermore, these analyses ignore the transformative potential of religion, its capacity to affect social and political change, the importance of the role of both local and international leadership (e.g. the case of the Catholic Church and Transnational Islam), issues of moral guidance and trust, and the way they relate to different socio-political and economic breakdowns. In short, normative political and developmental analyses of this sort often do not take into account the fact that religions are not static and that they constantly evolve and interact with contexts.

Religion and development

Development and modernity are intimately related to one another. Contemporary critiques of development theory assert that 'development' poses solutions to development problems in a peculiarly apolitical, antiseptic, neutral way (Escobar, 1995). The idea of religion, of placing 'unreasonable faith' above 'enlightened rationality', does not fit comfortably within development narratives that revolve around 'linear progressions and optimistic teleologies' (Ferguson, 1999, p. 13). This is largely due to a mix of fear and suspicion on the part of secular institutions and secular people who have to deal with a sphere (the spiritual) that does not seem to belong to the real. However, even within development discourses we often deal with principles like capital, debt and social structures that are invisible realities and vacuous entities.[1] Does secular development itself not invoke systems of beliefs (like progress) and defer to the existence of holy authorities and sacred spaces (like the World Bank!)?

If, as Pieterse states (2000, p. 182), development involves 'telling other people what to do in the name of modernisation, nation-building, progress [...] poverty alleviation and even empowerment and participation' we can be confident that at least a healthy portion of development is primarily concerned with telling other people what they ought to be thinking. We can equally confidently assume that this preferred mode of thinking is not likely to identify religious thought as a priority.[2] A large body of literature suggests that development has not marched non-Western countries towards a state of 'modernity'[3] in quite the way it was anticipated, promised or hoped for (Crush, 1995; Escobar, 1995; Ferguson, 1999; Power, 2003). Indeed, Ferguson, in his study of the decline of the Zambian Copperbelt, creates an image of Zambians watching development and modernity recede back over the horizon as the price of copper collapses (Ferguson, 1999). We need to problematize modernity and delineate 'development'. Religious organizations have often picked up the slack precipitated by failures of the state and short-comings in development (Myers, 1999; Kliksberg, 2003; World Faiths Development Dialogue, 2003; World Vision International, 2003). Our reading of this situation suggests that the Western notion of modernity as applied in non-Western contexts to underpin development has failed on two counts. First, as prevailing post-development received wisdom suggests, the types of development posited by notions of the modern have not been successful. Secondly, modernization assumed that the non-secular would gradually recede from public life when in reality the non-secular is pervading the spaces that the secular has singularly failed to fill. These perspectives are central threads running through the volume.

Development in the twentieth century Western European model was seen as progress from a deeply religious, irrational and non-bureaucratic world, to a modern space in which material advancement was achieved through 'sound' bureaucratic structures which led to secularization and the loss of the spiritual. This trajectory has not been universal and the United States of America is a good example of a developed nation-state that has not lost all sense of the spiritual and religious discourse in the public (Stout, 2004). Here we need to critique the idea that development inevitably leads to modernity and secularism. Should Africa be 'developing' along some pre-defined bureaucratic, secular pathway to modernity? Or as Comaroff and Comaroff argue, are there different pathways that lead to different models of modernity within a neo-liberal global economy (1993, p. ix). As we begin to think of development not as something that will supersede religion but rather interact with it in complex ways new questions must be posed: How do different religions define and critique development or understand development? Or, how is development shaped by religion or religious movements/communities (re)shaped by development? It is important to understand the shift from progress to process in understanding development. Progress is about material increase while process tries to bring about the advancement of the whole person and their society. Development should be about helping people realize their potential – what they are, what they can do for themselves, and who they could be (Sen, 1999). Since the economic crash in 2008 new discussions about the centrality of material gain as the measurement of human achievement have emerged, and this gives new impetus to the broader notion of the all-rounded individual which

Amartya Sen argues we should be striving for (1999). The focus should be on 'human development' and not only 'economic development', and 'development' itself should be more aligned to personal transformation, which as a by-product will bring about material improvement.

For academics like James Ferguson (1990) and Arturo Escobar (1995) development as a vehicle of modernity is almost the ultimate expression of power, drafting compelling representations of how the world ought to be. Politics, too, is bound up in those who can wield power and resources and those who cannot. Religion has long given people the possibility to speak to power, and through religious organizations the wherewithal to engage with existing power structures. Belief systems allow us to make sense of fields of power and the structures that recreate them. Through interpretation, healing, allegory and social action these institutions can equip communities with new ways to engage, contest and communicate. Religion counteracts power through the creation of power amongst other things, and politics or development in this context, can no longer be seen as unitary monoliths of power, purpose and design – but rather as spheres to be engaged with. This poses a challenge for development and politics, people move between different powers and leaders and ideas; there is not one power or single influence. Development organizations and political elites can be seen as a form of power to be engaged with among many others. Within this context the Western idea of a unique teleology of development does not seem to work effectively.

Religion and politics

Although in many African cultures a continuum exists between visible and invisible worlds (Appiah, 1992) and religion reaches all sectors of public life, Western analysts and disciplines generally struggle with considering how religious ideas come to have a bearing on the way political power is actually perceived and exercised (Haynes, 1994, p. 103). The wide spread of religious trends, common to other continents such as Latin America (Corten & Marshall-Fratani, 2001, p. 15), is characterized in Africa by the increasing role of religious actors in politics.[4] From the end of the 1980s, with the collapse of the one-party system and trends towards democratization, there was a concurrent erosion of legitimacy in African nation-states. In this period Christian churches, for example, played a remarkable role in preventing political *impasse*, in criticizing threats to the democratization process, and in promoting democratic values whilst acting inside civil society (Gifford, 1995, 1998). With the liberalization that took place during the 1990s new forms of Islamic organizations for youth, women and students became active participants in civil society and in the last 20 years public Islam has had a more pronounced public presence in Africa (Eickelman & Piscatori, 1996; Soares & Otayek, 2007). Steven Ellis and Gerrie ter Haar (Ellis & ter Haar, 1998) rightly affirm that the francophone school of political science led by Jean-François Bayart, strongly influenced by wider literature of philosophy, history and anthropology, has succeeded rather better than the anglophone tradition of political science in incorporating religion into its frame of analysis (Mbembe, 1985; Bayart, 1993; Mbembe & Toulabor, 1997; Mbembe, 2002).

In the new orders of the post-colony and of pluralistic societies, where one possible answer does not exist, the rise of the new religiosities and the *reassemblage* of ancestral voices have provided answers to the needs of these new orders among the numerous possibilities in social and political contexts (Mbembe, 1985). These values, as part of a process of identity reconstruction, are particularly important in Africa because they challenge the state in the utilization of 'resources of extraversion', to paraphrase Jean-François Bayart (Bayart, 1993) and because they elaborate an alternative vision of modernity and future. Religious and cultural phenomena are not free from changes embedded inside this process of reinterpretation, as they are historical constructions of it. Democracy, too, has different meanings for different actors inside society. It is important to analyse changes in religious ideas and for religious people within the broader transformations of nation-building and how they interact with other social agents. The ability of religion to deal with the challenges of the post-colonial period and rapid social changes could be one of the sociological explanations of the rise of Christianity and Islam in Africa in the last few decades (Isichei, 1995; Levtzion & Powells, 2000).

In the wake of a new school of thinkers on development in Africa, we can observe a new school of political scientists who are involved in a process of re-evaluation of the meaning of religious and so-called *non-*modern (or *non-*Western) values inside the socio-political institutions of the continent. Achille Mbembe, for example, accused political science and development economics of having undermined 'the very possibility of understanding African economic and political facts' (Mbembe, 2001, p. 7). Religion can be a potent agent able to mobilize people and synergies in a profound way, while to ignore religion, as a matter of obvious political and even economic importance, threatens the credibility of academic investigations.

We believe it is important to analyse changes in religion itself and its interactions with other social actors and political dispensation. It is on the continuum between private and public, and between the individual and the community, that we need to focus our attention to understand local interpretations of religion. If it is important to understand the public role of religious organizations, of leadership and public voices, it is also prudent to understand the relevance of the spiritual power of religion, of non-organized groups, and movements who do not access mass media, ideas and beliefs, but that may bring about social change. This is about understanding social shifts on the ground and the force of power from below within the broad public arena.

Religious spaces in the African state

Literature from the 1960s and 1970s that dealt with religion in Africa continues to dominate contemporary investigations of religion in the continent, giving a sense that there was little left to study, and little need to do so, as Africa was inevitably going to follow a Western path of modernization. So powerful was this discourse that recent analyses highlighting the enduring reality of religion have simply been portrayed as a negation, or a deviation, from a so-called Western normality (Mbembe, 2001). In most debates about Africa the interaction between the three areas of religion, politics and development has not been studied in depth. It is vital, therefore, to bring these three factors into dialogue with one another, as a reflection of social realities. The general relevance of a focus on religion is that lifeworlds in much of sub-Saharan Africa are paradigmatically shaped by religious frameworks.

The rise of religiosity in the Global South, especially through Pentecostalism and the active proselytism of international Islamic groups in West and East Africa, has forced a rethink of the role of religion in the public sphere; but this is not the only phenomenon we should consider. Africa faces a challenging twenty-first century, as perhaps we all do, but issues such as climate change, food insecurity, deepening economic disparities, health and illness, and conflict will in all likelihood be most keenly felt there. We should not forget that the context in which these problems are to be solved will endure as an African one. Bruce Janz (2004) makes the compelling argument that there is such a thing as an 'African philosophy' because any thought process, identity formation or cultural symbolic system is shaped by the history, ideas, forms of power, community formations and environment specific to a geographical space and so form the particular contexts in which specific ideas, philosophies and ways of forming meaning may emerge. Mbembe (2004) argues that if we are to fix or understand, or truly sympathize with, the multiple difficulties faced by people living in Africa we have to do so by finding African ways of thinking about, articulating and maybe even solving some of the problems. This volume is an attempt to speak about issues of development, religion and politics from an African perspective, from the bottom up, from the people who live and die in the broader geographical and historical context of Africa (Chabal, 2009).

Ellis and ter Haar (2004) have explained how the forces of religion and politics are continuously in dialogue with one another and when we separate them out we create a simplistic and therefore unhelpful picture. We believe it is important to include the dynamics of development into this debate. There has been little opportunity to bring together a comprehensive comparative analysis of the interactions between religion, politics and development in Africa. In this book we have aimed to challenge this lack of interaction and create a dialogue between the three elements. We hope to offer a framework within which to draw together a meaningful analysis of these interactions, as well as offer a broad perspective of historical and contemporary empirical material which can inform this analysis.

Overview and content of the book

This edited volume represents the outcome of the 2008 conference 'Exploring Religious Spaces in the African State. Development and Politics from Below'.[5] The first section, 'Challenging the Secular: Religion and Public Spaces', focuses on the important role that religion has played in shaping the public spaces in different parts of Africa. In the first chapter, 'Development and Invisible Worlds', Stephen Ellis gives an overview of the history of development studies in Africa. He argues that in Africa there is not the same division between the sacred and the secular as there is in countries dominated by Western philosophical ideas of sacred verses spiritual. Throughout Africa there is instead a 'division' between the seen and the unseen, but both worlds are infused with the spiritual, and the manifestation of spiritual powers within the political and public arena is regarded as a given fact. Ellis shows that in much of the secular development work done in Africa through the twentieth century the importance of religion was discounted and development regarded as the way towards modernity. This modernity was understood as leading to bureaucratization, a fading away of all that was spiritual and a focus on material wealth through capitalist economic structures.

This approach has had limited success partly because it does not take into consideration the *Weltanschauung* of people living in Africa.

In the following chapter, 'The Mbuliuli Principle: What is in a Name?', Gerrie ter Haar asks the important questions, what if we brought religion more fully into development analysis? What should development be about and what should it achieve? To answer these she examines the important role that religions as meaning systems – ways of understanding the world and the self – play in Africa. Having laid this ground she then explores the economic principle for development first established in the mid-1970s by Emmanuel Milingo, who was then the Catholic Archbishop of Lusaka. Archbishop Milingo was deeply involved in a ministry that tried to bring about healing to the whole person and society. He supported reading programmes, social upliftment causes, and intervention that worked towards physical, emotional and spiritual religious healing. In his work he was outspoken in his criticism of African governments whom he recognized as corrupt and was passionate about helping his people recognize their own potentials for economic independence. Milingo developed what he called the Mubliuli Economic Principle, which was an economic/self-development programme based on the corn seed, which, when exposed to intense heat pops open to become popcorn. Milingo's principle was that people should be like corn seed and use their situations of intense heat that is, their difficulties, to think differently and understand their innate abilities, which will enable them to become profitable, whole people. Working with Milingo's ideas ter Haar challenges development programmes to move away from progress thinking, which is focused primarily on material upliftment, to seeing people as whole beings with social, emotional, spiritual and material needs. She argues for a new type of development which is process driven and seeks to help the whole person become transformed through a process whereby they can realize their own sense of agency and personal potential. Within this process there is also a strong social dimension, which can enable people to become trusting of their communities again, develop healthy relationships and become aware of their own human dignity and that of those around them.

The ideas of a social world based on human dignity and freedom lies at the heart of the next chapter in this section, 'Muslim Shrines in Cape Town: Religion and Post-Apartheid Public Spheres', in which Abdulkader Tayob explores the importance of Muslim burial shrines (*kramats*) around Cape Town in South Africa. Tayob gives an in-depth study of the legal and religious drama which erupted on to the political landscape of Cape Town between 1996 and 2007, and shows how these shrines reveal the complex engagement of culture freedom in post-apartheid South Africa. The *kramats* are the sites where Muslim saints, dating back to the eighteenth century, were buried, and as such they are sacred spaces within the Muslim geography of South Africa. During the apartheid era permission was granted for the land on which the *kramats* are located to be developed into a housing estate. For various reasons the development did not take place and in 1997 the owners of the land again logged a request for a housing permit to develop this land. The case was taken to court and sparked a passionate debate within the Muslim community about religion, the place of the *kramats* and the changing nature of contemporary Muslim identities and worship. This insightful account highlights the very public nature of religion in a country like South Africa which has tried to become secular, at least on a political level. Questions about the shape of Muslim identity are raised and the whole discussion of this chapter shows how religious interests can and must at times take precedence over economic development.

The importance of religion in personal transformation and social upliftment is also highlighted by Dorothea Schulz in her chapter, 'Remaking Society from Within: Extraversion and the Social Forms of Female Muslim Activism in Urban Mali'. Schulz analysed why women in Mali – particularly in and around the capital Bamako, have chosen to express their religiosity so much more openly in the last few years. What she found in Mali were new forms of piety and objective representations of female morality in the public domain that were expressed in new and renewed configurations of social and political alliances which were often influenced by Western, donor-supported 'development' structures and discourses.

> 'One' important difference between these groups and their historical predecessors is that their social basis and age composition has changed. Nowadays, many participants of Muslim neighbourhood groups and networks come from the urban middle and lower-middle classes and represent a younger generation of married women. The organisational structures of the groups and the objectives of their gatherings have partly changed, too.
>
> (Schulz, 2010)

Their commonality before Allah makes all these women equal yet their educational, social, economic and political status outside religious institutions continues to inform these groups so that different hierarchical systems exist within the 'equality'. And much of the capital that these women have acquired has been through the input of foreign donor-funded programmes that have given them new skills and new views which they bring into their public and religious life. With their new voices and tools of debate these Muslim women have become activists in Mali, engaged in a particular modality of 'politics from below' where they are not trying to bring about or voice political protest, but to transform the personal and social reality of all Malians.

The second section of this volume looks at how religions function, inform and shape communities and people in that space found between the state and society. David Skinner's chapter '*Da`wah* in West Africa: Muslim *Jama`at* and Non-Governmental Organisations in the Gambia, Ghana and Sierra Leone, 1960–90', is the first in this section 'Religion Between State and Society'. In this chapter Skinner focuses on

> the creation and maintenance of Islamic space and efforts by Muslims to expand their political, economic and social influence in these states through the formation of nongovernmental organisations and their interaction with governments and international agencies.
>
> (Skinner, 2010)

He looks at how Islam has grown rapidly in West African countries and has begun to influence some of the broader political spheres in Ghana, The Gambia and Sierra Leone. On the whole, he argues that NGOs, and particularly Muslim NGOs, are interested in expanding their own development agendas which are not always related to that of the government nor do they always work in consultation with government bodies. While there are many Muslim NGOs working in these countries they do not have a co-ordinated plan of delivery and their communication with one another is poor, largely because of the different ethnic communities who make up the Muslim community throughout these three countries. This lack of communication makes these Islamic NGOs less effective than they could otherwise be and weakens the political voice they might have. In all three states the Muslim NGOs have been filling the gaps left by states that do not provide adequate social and economic development programmes. The Muslim organizations have found themselves competing with each other for funding, a political voice and international recognition within the Muslim community. Two primary public functions filled by Muslim organizations in all three countries are health care and education. Their influence is so extensive that many young people receive most if not all of their education through Muslim run and funded schools. Through Islamization this part of Africa has become directly linked into the Middle Eastern conflict in that much of the donor aid received by Muslim NGOs comes from Middle Eastern countries – particularly Kuwait and Saudi Arabia. The continued development of these countries has therefore become more directly linked into global economic realities as donor funding is dependent on the ability of other Muslim countries to give to their African brothers.

The importance of faith based organizations (FBOs) is examined in another part of Africa by Ernest Mallya, who looks at FBOs in Tanzania in his chapter, 'FBOs, the State and Politics in Tanzania'. Like Skinner, he shows that the FBOs in this country also fill the gap left by impoverished and corrupt post-colonial governments who have been unable or unwilling to offer the social and material resources needed for Tanzania. This chapter offers important insight into the tenuous relationship between NGOs, FBOs and government in an African state. He suggests that the Tanzanian government, which prides itselfon being secular, has an ambivalent relationship to FBOs – using them when they are able to fill the political gap of service delivery and resources when they cannot, and immediately limiting their sphere of influence, even to the point of taking over all their assets, when they have the resources or political will to do so. This has meant that FBOs in Tanzania have been limited in what they are able to achieve. As their 'progress' is weakened by the interference of government organizations and there is no guarantee that in the future the government will not take over hospitals, schools and other infrastructures set up by FBOs, these organizations struggle to secure the international funding they require to effectively run projects that will fill the gap left by the state. In both Skinner's and Mallya's chapters the gap between state and society is clearly shown and in different ways the authors analyse how this gap is being filled by religious organizations. The social and political result of the involvement of religious organization has a variety of outcomes, both positive and negative.

The final chapter in this section, by Linda van de Kamp, examines the space between state and society from a different angle. In her chapter 'Burying Life: Pentecostal Religion and Development in Urban Mozambique', van de Kamp offers a critique of the wider development practices of the Universal Church of Christ in Mozambique. This chapter exposes the religious practices of this Pentecostal Charismatic Christian Church with specific regard to its economic teachings. Van de Kamp found that one of the most important rituals within this church was that of sacrificing money to the church. People would give excessive amounts of their own money to the church as an offering to God, appealing to him to bless them with wealth. The result of these offerings was that pastors often became wealthy while people became poorer, even bankrupt.

> In Mozambique (where the study was conducted) many converts do not gaze at consumer items, but sell and buy them. They partake in the capitalist economy. They work in companies, banks, NGOs, and run (small) shops. They do not find the neoliberal order mysterious.
>
> (van de Kamp, 2010)

Why then do these informed people offer so much money to their churches? What these pastors are offering their congregations are ways in which to realize their personal potential and to be liberated from the demands of their kinship networks. This form of upliftment, which seeks to develop the whole person, is paradoxically manifest in ritual techniques that produce socio-cultural discontinuities, risks and vulnerability. The religious, social and psychological teaching of these churches enables people to make social, psychological and spiritual breaks with their kinship networks and traditional philosophical and religious mindsets (Mbiti, 1991). This should enable people to more freely embrace capitalistic and individualistic forms of modernity; but for many people it becomes a road to further brokenness and increased vulnerability. What this chapter highlights explicitly and what the previous two chapter at times allude to is that religious institutions and FBOs may not always have positive influences and might not always lead people to a path of transformation and emancipation – in practice they can lead to greater dependence and poverty.

Critically evaluating what a religious meaning system might have to offer an individual person or community lies at the centre of the debate raised in the final section of this book, 'Health Care Provision: Reflections on Religion'. This section begins with a theoretically challenging chapter, 'Health and the Uses of Religion: Recovering the Political Proper?', by James Cochrane. Drawing on research into religion and health Cochrane asks whether religions might offer the dynamics necessary to recover 'the political proper',

> that discursively redeemed space of the public, currently threatened, in which the communicative interaction necessary to subdue and tame the destructive excesses of political power and the market economy becomes possible.
>
> (Cochrane, 2010)

Working with Habermas's idea of lifeworlds he argues that we have different spheres of human life, one of which is the lifeworld of health, which people understand in different ways. Religion may be an asset in the lifeworld of healing, in other words it might be a way, a system of thought, in which health is understood and healing undertaken differently from the bio-medical models of healing normally proffered by development workers informed by Western medical technologies. Working with religious leaders rooted in different traditions Cochrane proposes that 'if one is concerned about influencing the structural formations of the state (polity) and the market (economy) in ways that reduce their deficits (including their erosion of prized lifeworlds)', we need to 'leverage those norms and values internal to religious identities that are inclusive, extensive and transformative' (Cochrane, 2010). This would mean working with multiple approaches gathered from different religious and cultural viewpoints to bring about holistic healing and advancement for all people living in Africa. Part of this process towards wholeness would be a reclaiming of the public sphere as something which is truly politic and in which citizens are actively engaged.

The next chapter in this section, 'Marshalling the Powers: The Challenge of Everyday Religion for Development', moves away from the more theoretical need for a public sphere and how religions and religious leaders might reopen this engagement to a view of health and religion from the perspective of rural villagers living in Ghana. This chapter, by Elisabeth Graveling, examines how people living in a very remote and poor part of Africa make informed decisions about the spiritual powers they understand to be around them all the time, and how they use different religions to manage these powers. For development work to truly become about a process towards healing, transformation, personal agency and independence it needs to take on board what people think and say, and understand how they use resources currently available to them, such as religion, to manage their own lifeworlds as best they can. Cochrane and Graveling ask us to see that in the lives of ordinary people in Africa the world does not exist in neat, discreet compartments of traditional religions, various forms of Christianity or Islam, but that these are all seen as part of the same landscape, a landscape of spiritual powers that need to be managed and for which differing languages, technologies and insights can have varying degrees of efficacy. Therefore it is entirely logical for people to go to the hospital, the sangoma and the imam to seek healing from a physical aliment.

Awareness of the complexity of people's worldviews and experiences with regard to health is discussed further by Ezra Chitando in his chapter, 'Sacred Struggles: The World Council of Churches and the HIV Epidemic in Africa'. Chitando shows how the World Council of Churches (WCC) has been involved in development work through its programmes to make people aware of the HIV/AIDS epidemic and pro-mote social acceptance of people who are either suffering from the disease or are affected by it. This, he argues, is a different type of devel-opment work from the traditional work of material improvement: it is a form of development that takes in consideration socio-cultural and health care issues. The review of the WCC offered by Chitando shows just how difficult working to alleviate the strain of HIV/AIDS in Africa is and how the WCC, while being well intentioned, has had limited suc-cess in getting the different member churches to implement HIV/AIDS prevention, intervention and care programmes that impact on com-munities in such a way that the epidemic becomes controlled. The thousands of volunteer workers trained through the WCC workshops go back to work in relative isolation in their communities, making it diffi-cult to bring about real change – change from the bottom up. The WCC programmes, like other FBO work, rely on funding from donors, and this funding is not always available. This section on healthcare, religion and the public space shows the important role that religions, in dif-ferent forms, can play in offering healing not only through traditional bio-medical models of mission hospital but also through the important social, spiritual and emotional healing which different religious bodies offer in different ways.

Notes

1. See Chapter 1, Ellis, S. 'Development and Invisible Worlds'.
2. This has certainly be the case historically although now the World Bank for example are beginning to identify churches as an important development tool, if not as important belief systems. See for example Pallas, C. (2005), 'Canterbury to Cameroon: A New Partnership Between Faiths and the World Bank', *Development in Practice*, 15 (5), pp. 677–684. Belshaw, D., Calderisi, R. & Sugden, C. (2001), *Faith in Development: Partnerships Between the World Bank and the Churches of Africa*, Washington, D.C.: World Bank.
3. We embrace Knaupt's idea of modernity: 'Modernity can be defined as the images and institutions associated with Western-style progress and develop-ment in a contemporary world' (Knaupt, 2004, p. 18).
4. This thesis can find support in Mbembe, 1985; Bayart et al., 1992; Chidester, 1992, 2001; Gifford, 1995, 1998; Haynes, 1996; Maxwell, 1999; Corten & Marshall-Frantani, 2001; Haynes, 2002; Soares & Otayek, 2007.
5. 'Exploring Religious Spaces in the African State. Development and Politics from Below', conference organised by the Centre of African Studies, the Uni-versity of Edinburgh in collaboration with WISER, Wits Institute for Social and

Economic Research, the University of the Witwatersrand, Edinburgh, 9–10 April 2008.

References

Allen, D. (ed.) (1992), *Religion and Political Conflict in South Asia: India, Pakistan and Sri Lanka*, London: Greenwood Press.

Alkire, S. (2006), 'Religion and Development' in Clark, D. A. (ed.), *The Elgar Companion to Development Studies*, Cheltenham: Edward Elgar.

Anderson, A. & Tang, E. (2005), *Asian and Pentecostal. The Charismatic Face of Christianity in Asia*, Oxford: Regnum Books International.

Appiah, K.A. (1992), *In My Father's House: Africa in the Philosophy of Culture*, Oxford: Oxford University Press.

Bayart, J. F. (1993), *The State in Africa: The Politics of the Belly*, London: Longman.

Bayart, J. F., Mbembe, A. & Toulabor, C. (eds) (1992), *La politique par le bas in Afrique noire*, Paris: Karthala.

Belshaw, D., Calderisi, R. & Sugden, C. (2001), *Faith in Development: Partnerships between the World Bank and the Churches of Africa*, Washington, D.C.: The World Bank.

Berger, J. (2003), 'Religious Nongovernmental Organizations: An Exploratory Analysis' *Voluntas: International Journal of Voluntary and Nonprofit Organizations*, 14 (1), pp. 15–39.

Berger, P. (ed.) (1999), *The Desecularization of the World: Resurgent Religion and World Politics*, Washington D.C.: Grand Rapids.

Brass, P. R. (2003), *The Production of Hindu-Muslim Violence in Contemporary India*, University of Washington Press: Seattle.

Casanova, J. (1994), *Public Religion in the Modern World*, Chicago: The University of Chicago Press.

Chabal, P. (2009), *The Politics of Suffering and Smiling*, London: Zed Books.

Chidester, D. (1992), *Religions of South Africa*, London: Routledge.

Chidester, D. (2001), *Christianity: A Global History*, London: Penguin.

Clarke, G. & Jennings, M. (eds) (2008), *Development, Civil Society and Faith-based Organizations*, London: Palgrave-MacMillan.

Comaroff, J. & Comaroff, J.L. (1993), 'Introduction' in Comaroff, J. & Comaroff, J. L. (eds), *Modernity and Its Malcontents: Ritual and Power in Postcolonial Africa*, Chicago: University of Chicago Press.

Corten, A. & Marshall-Fratani, R. (2001), *Between Babel and Pentecost: Transnational Pentecostalism in Africa and Latin America*, Bloomington: Indiana University Press.

Crush, J. (eds) (1995), *Power of Development*, London: Routledge

Davies, G. (2002), *Europe: The Exceptional Case. Parameters of Faith in the Modern World*, London: Darton, Longman and Todd.

Deneulin, S. & Bano, M. (2009), *Religion in Development: Rewriting the Secular Script*, London: Zed Books.

Ellis, S. & ter Haar, G. (1998), 'Religion and Politics in Sub-Saharan Africa' *Journal of Modern African Studies*, 36 (2), pp. 175–201.

Ellis, S. & ter Haar, G. (2004), *The Worlds of Power: Religious Thought and Political Practise in Africa*, Oxford: Oxford University Press.

Eickelman, D. F. & Piscatori, J. W. (eds) (1996), *Muslim Politics*, Princeton: Princeton University Press.

Escobar, A. (1995), *Encountering Development. The Making and Un-making of the Third World*, Princeton University Press: Princeton.

Ferguson, J. (1990), *The Anti-Politics Machine. Development, Depoliticisation and Bureaucratic Power in Lesotho*, Minneapolis: University of Minnesota Press.

Ferguson, J. (1999), *Expectations of Modernity. Myths and Meaning of Urban Life on the Zambian Copperbelt*, Berkeley: University of California Press.

Fetzer, J. S. & Soper, J. C. (2005), *Muslims and the State in Britain, France, and Germany*, Cambridge: Cambridge University Press.

Gifford, P. (1995) (eds), *The Christian Churches and the Democratisation of Africa*, Leiden: Brill Press.

Gifford, P. (1998), *African Christianity and its Public Role*, London: Hurst and Company.

Haghayeghy, M. (1996), *Islam and Politics in Central Asia*, New York: St Martin's Press.

Hansen, T. B. (1999), *The Saffron Wave: Democracy and Hindu Nationalism in Modern India*, Oxford: Oxford University Press.

Haynes, J. (1994), *Religion in Third World Politics*, Buckingham: Open University Press.

Haynes, J. (1996), *Religion and Politics in Africa*, London: Zed Books.

Haynes, J. (2002), *Politics in the Developing World: A Concise Introduction*, London: Blackwell.

Haynes, J. (2007), *Religion and Development: Conflict or Cooperation?* London: Palgrave-MacMillan.

Heelas, P. (ed.) (1998), *Religion, Modernity and Post-Modernity*, Oxford: Blackwell.

Isichei, E. (1995), *A History of Christianity in Africa. From Antiquity to the Present*, London: Society for Promoting Christian Knowledge.

Janz, B. (2004), 'Philosophy As If Place Mattered: The Situation of African Philosophy' in Carel, H. & Gamez, D. (eds), *What Philosophy Is: Contemporary Philosophy in Action*, London: Continuum International Publishing Group.

Juergensmeyer, M. (2000), *Terror in the Mind of God: The Global Rise of Religious Violence*, Berkeley: University of California Press.

Kliksberg, B. (2003), 'Facing the Inequalities of Development: Some Lessons from Judaism and Christianity' *Development*, 46 (4), pp. 57–63.

Knaupt, B. M. (ed.) (2004), *Critically Modern. Alternatives, Alterities, Anthropologies*, Bloomington and Indianapolis: Indiana University Press.

Levtzion, N. & Powells, R. L. (eds) (2000), *The History of Islam in Africa*, London: James Currey.

Mamdani, M. (2004), *Good Muslim, Bad Muslim: America, the Cold War, and the Roots of Terror*, New York: Pantheon Books.

Marshall, K. & Keough, L. (2004), *Mind, Heart and Soul in the Fight against Poverty*, Washington, D.C.: The World Bank.

Marshall, K. & Van Saanen, M. (2007), *Development and Faith: Where Mind, Heart and Soul Work Together*, Washington, D.C.: The World Bank.

Martin, D. (1990), *Tongues of Fire*, London: Blackwell.

Martin, D. (2002), *Pentecostalism. The World their Parish*, London: Blackwell.

Maxwell, D. (1999), *Christians and Chiefs in Zimbabwe*, Edinburgh: Edinburgh University Press.

Mbembe, A. (1985), *Afrique Indociles. Chistianisme, pouvoir et État en société postcoloniale*, Paris: Ed Karthala.

Mbembe, A. (2001), *On the Postcolony*, Berkley: California University Press.

Mbembe, A. (2004), 'Aesthetics of Superfluity' *Public Culture*, 16 (3), pp. 373–405.

Mbembe, A. & Toulabor, C. (1997), *Religion et Transition Democratique en Africque*, Paris: Ed Karthala.

Mbiti, J. S. (1991), *Introduction to African Religion*, 2nd edn, Oxford: Heinemann Educational Books Ltd.

Myers, B. L. (1999), *Walking with the Poor: Principles and Practices of Transformational Development*, Marynoll: Orbis/World Development.

Norris, P. & Inglehart, R. (2004), *Sacred and Secular: Religion and Politics Worldwide*, Cambridge: Cambridge University Press.

Pallas, C. (2005), 'Canterbury to Cameroon: A New Partnership Between Faiths and the World Bank' *Development in Practice*, 15 (5), pp. 677–684.

Peterson, A. (1996) 'Religion and Society in Latin America: Ambivalences and Advances' *Latin American Research Review*, 31 (2), pp. 236–251.

Pieterse, J. N. (2000), 'After Post-development' *Third World Quarterly*, 21 (2), pp. 171–191.

Power, M. (2003), *Rethinking Development Geographies*, London: Rutledge.

Rist, G. (2002), *The History of Development. From Western Origins to Global Faith*, London: Zed Books.

Romero, C. (2001), 'Globalization, Civil Society and Religion from a Latina American Standpoint' *Sociology of Religion*, 62 (4), pp. 475–490.

Roy, O. (2004), *Globalized Islam, the Search for a New Ummah*, New York: Columbia University Press.

Roy, O. (2007), *Secularism Confronts Islam*, New York: Columbia University Press.

Sen, A. (1999), *Development as Freedom*, Oxford: OUP Oxford.

Soares, B. & Otayek, R. (2007), *Islam and Muslim Politics in Africa*, London: Palgrave-Macmillan.

Stoll, D. (1990), *Is Latin America Turning Protestant? The Politics of Evangelic Growth*, Berkley: California University Press.

Stout, J. (2004), *Democracy and Tradition*, Princeton: Princeton University Press.

Tyndale, W. R. (ed.) (2006), *Visions of Development: Faith-Based Initiatives*, Aldershot: Ashgate.

World Faiths Development Dialogue (2003), 'The Provision of Services for Poor People: A Contribution to WDR 2004', mimeo, electronic resource: www.wfdd.org.uk. [accessed 10 October 2009].

World Vision International (2003), 'Annual Report', electronic resource: http://www.worldvision.com.au/AboutUs/CorporateGovernance/AnnualReports.aspx. [accessed 10 October 2009].

Part I

'Challenging the Secular: Religion and Public Spaces'

1
Development and Invisible Worlds

Stephen Ellis

A notion of development, however vague, is implicit in the proposition that 'human beings can act, collectively, to improve their lot' (Leys, 1996, p. 3). Some writers believe that the concept that human beings are capable of collective self-improvement was held even in antiquity (Rist, 2002), while others suggest that it was the rise of industrial capitalism that brought the fact of human development forcibly to people's attention for the first time, enabling Hegel and Marx to see world history as a process of development (Leys, 1996, p. 4).

In this light, it is not difficult to see why religion has sometimes been regarded as an obstacle to development. Since many varieties of religion postulate the existence of spiritual forces and spiritual beings that are deemed to be independent of human control, in the minds of believers these same entities may constitute objective limits to the ability of people to act collectively to improve their condition. A view of human history and human capacities that is founded on secular and materialist assumptions, emphasizing the ability of mankind to shape the world to its own design, may thus appear to be inherently opposed to religion.

Yet development and religion are more closely related than it would at first appear. Development, as it has been understood in recent times, is more than a broad conviction that human beings can do something to improve their lives, rather it is part of a fairly specific set of ideas about the world that is rooted in Europe's historical experience. It is when this particular historical path is examined in more detail that the relationship between religion and development becomes clearer. In particular, it becomes apparent that even the secular, technocratic idea of development that has been predominant in recent decades has connections to the Christian ideas of cosmology that held sway in Europe for so many centuries. According to the Indian economist Deepak Lal (1998, p. 177), the intellectual progenitor of the idea of progress through history that is essential to contemporary ideas of development was none other than the great Christian theologian Augustine of Hippo. At the heart of Augustine's theology is the idea that history is a story whose meaning unfolds as it builds up to a great climax in the form of the millennium and the return of Christ to earth, foreseen in the final book of the Bible. Over centuries, this apocalypse has been secularized and turned into a theory of permanent growth, with the millennium now translated into a vision of a world in which everyone can live a fulfilled life free from hunger and disease. The pursuit of the Christian millennium (cf. Cohn, 1957), one might say, has turned into the pursuit of the millennium goals.

In short, development as it has actually emerged, as an intellectual sub-discipline and a bureaucratic practice, is to a great extent the offspring not just of religion, but of a particular religion – the Christianity that evolved in Western Europe from the last period of the Roman empire. The story of how a religious vision of the cosmos gave way to a materialist goal of a world of abundance is inseparable from the lessons learned from Europe's past. For at least two centuries, writers steeped in a heroic vision of Europe's achievements have extrapolated these from European data to produce itineraries for the future of humankind. The Reformation, the rise of the secular state, the Enlightenment and the French Revolution are the stuff of a history so widely disseminated as to have achieved the status of common knowledge. The process of sorting through Europe's history to draw universal conclusions has also shaped social science, which, in its origins, was based on neither more nor less than a reading of European history conceived in scientific mode, by thinkers who believed they could identify, on the basis of these data, general laws concerning human behaviour.

There are now compelling reasons to reconsider the road-maps that have formed the intellectual basis for development studies and development policies, whose distant origin lies in Christian theology. First and foremost, it has become evident that other historical paths can be taken by societies that are intent on increasing their command over resources, as Asia has shown. Asian countries did not achieve their current status by emulating Europe in every respect, or even in all key respects. Realizing the salience of some new narratives of improvement, increasing numbers of historians are inclined to re-examine the historical period during which the West achieved such extraordinary hegemony, avoiding the assumption that Europeans were the only historical actors whose ideas and actions mattered in the sense of having lasting influence on the world (Bayly, 2004).

It is therefore necessary to note the variety of ways in which people may conceive of improvement, and the equal diversity in how visions of material abundance may be translated into lived reality. A society may be perceived by its members as having the capacity to become more affluent not only in the sense of exercising control over physical objects, but also in enriching its endowment in non-material resources, including the psychological and spiritual resources that may affect such intangible assets as security, satisfaction and happiness.

> Human Development is a development paradigm that is about much more than the rise or fall of national incomes. It is about creating an environment in which people can develop their full potential and lead productive, creative lives in accord with their needs and interests' the United Nations asserts.
>
> (UNDP, 2009)

'Development is thus about expanding the choices people have to lead lives that they value. And it is thus about much more than economic growth, which is only a means – if a very important one – of enlarging people's choices' (ibid.). This broad view of human development is relatively recent, however. In earlier decades, when development was largely synonymous with modernization, a much greater emphasis was placed on economic growth and on the adoption of Western-style institutions than on choice.

If we are to understand the meaning of development as a lived experience, we must consider how it has historically been applied. In the case of Africa, large numbers of people since the mid-twentieth century have witnessed the rise of a formidable development industry founded on a technocratic endeavour to improve the human condition through bureaucratic activity; in which, development has tended to become equated with economic improvement in the first instance. Throughout this period, development has generally been considered by experts to involve a separation of religion from the realm of material improvement. Yet Africans have never ceased to regard material factors partly through the prism of religion, attempting to bring all the forces that shape their lives under human control.

Accordingly the present chapter will consider how technocratically inspired projects of development have affected Africans people's sense of control of their own destiny. In order to do this, it will first briefly discuss how European and African ideas concerning cosmology and religion have changed over time, and how a secular quasi-science of development has been received with a religious frame of thought.

Histories of African improvement

For centuries European contacts with Africa, south of the Sahara have been marked by an aspiration to change and to improve for centuries. When Portuguese seafarers began to trade in areas south of the Sahara in the fifteenth century, their government and their Church also embarked on a campaign of evangelization whose greatest triumph was the conversion of the king of Kongo, whose domain was in present-day Angola and the Democratic Republic of Congo. Subsequently, Africa's Atlantic coast, from the Senegal River to Cape Town, became absorbed into a system of maritime commerce, dominated by the slave trade, which was controlled by European naval powers. The continent's eastern coast, meanwhile, had for centuries been integrated into transcontinental systems of commerce, via trade routes that were dominated first by Austronesian navigators, later by Arab and Swahili skippers and traders, and only later still by their European competitors. In the Indian Ocean as in the Atlantic, the shipment of slaves from Africa to markets overseas was a staple trade. The interior regions of Africa, out of reach to seaborne traders who rarely travelled far inland, were familiar with localized patterns of exchange that in time became linked to coastal ports or across the Sahara via intermediaries. Nevertheless, the missionary impulse of the age of the Reformation and the Counter-Reformation was not accompanied by the aspiration to economic transformation that was to become so central to later notions of development.

A second great wave of European Christian proselytization arose in the late eighteenth century, this time led by British evangelicals, who founded the London Missionary Society in 1795. The British evangelical revival was also the nursery of the struggle to abolish the slave trade, perhaps the first great campaign to mobilize public opinion in something like its modern sense. The dynamic campaign of evangelization in Africa was subsequently joined by Protestants from other parts of the Europe and the United States, as well as by a new generation of Catholic missionaries. Ever since then, European and North American relations with Africa have not ceased to be marked by a moral dimension that originated in the slave trade and the campaign for its abolition.

This second great wave of European missionary activity in Africa, unlike the first, was associated with an ambition to effect an economic transformation. Early nineteenth-century Christian missionaries in Africa as well as many secular officials and travellers of that period expressed the conviction that they were bringing to Africa not only Christianity but also what they themselves termed 'civilization' (e.g. Ellis, 1838, p. 82), a word designating a transfer of European technical expertise and Christian religion. In the case of British missionaries, the civilizing mission was often coupled with an aspiration to bring Africa into the world of capitalist free trade, reflecting a contemporary view that the introduction of a monetary economy would transform society by creating markets. In Victorian times, the mission to civilize that emerged in Northern Europe in the late eighteenth century and was married to a Darwinian concept of racial science, was a direct forerunner of the modern idea of development.

In the early centuries of European exchange with Africa, Europeans just as much as their African partners thought about the world largely in religious terms. The European writers of those times generally considered there to be four types of religious practitioner in the world: Christians, Jews, Mohammedans (as Muslims were then known) and 'the rest', meaning all others deemed to be attached to some form of idolatry (Masuzawa, 2007, p. 181). Cultural contact with areas outside the world previously known to them caused Europeans to rework their ideas about both their own religion, Christianity, and about the practices and beliefs of others concerning the invisible world (Pagden, 1986; Asad, 1993). During their remarkable commercial, military and political expansion to bring the whole globe within the scope of their influence, Europeans came to consider themselves as occupying a higher stage of historical evolution than people in other parts of the world. They conceived of their superiority as both moral and technical. Over centuries, an externally conceived aspiration to transform Africa has become inseparable from several other projects or trajectories, including an evangelizing impulse, the expansion of capitalist markets and contests for global power. The urge to transform has spread from Western Europe to other places, notably North America.

The idea of a mission to civilize parts of the world that were perceived as stuck in time became an integral part of colonialism. The Berlin conference of 1884–85 accorded to rival European powers their own spheres of influence, which they set out to turn into effective rule in the manner Europeans best understood, by implementing a state monopoly of violence and by marking the exact frontiers of their territories. In the first stage of the colonial period, metropolitan governments made it apparent to their colonial officials that they were not much interested in using tax-payers' money to fund colonial infrastructure projects or improvement schemes. Later, during the great depression of the 1930s in both Britain and France there were tentative suggestions for using capital from the metropole in order to enhance the economies of colonial territories. This led, for example, to Britain's Colonial Development and Welfare Act of 1940 (Cooper, 2002, pp. 30–1), but such changes were small-scale and had limited effect.

It was, above all, in the aftermath of the Second World War and the emergence of the United States as a superpower that a broad aspiration to improve Africa evolved into the modern, technocratic form of development. US policy-makers developed a strategy of global economic expansion that was intended both to outflank Communism and to avoid the economic depression that had ensued after the First World War. President Harry Truman made this strategy a central plank of his second term in office, announcing in his inaugural address of 20 January 1949 that 'we must embark on a bold new program for making the benefits of our scientific advances and industrial progress available for the improvement and growth of underdeveloped areas'.[1] The American vision of a more prosperous world included the establishment of sovereign states that would replace the existing colonial empires. Meanwhile, the other superpower of the mid-twentieth century, the Soviet Union, enthusiastically endorsed the same principle of universal political sovereignty and economic progress as long as its own official myth – to the effect that it was a union of emancipated peoples rather than a Russian empire – was not questioned. In time, the Soviet government came to consider hegemony over the third world as the key to a global victory in the Cold War (Andrew & Mitrohkin, 2005, pp. 480–1).

The articulation of a strategy of development caused colonial administrations in Africa to introduce new policies intended to create economic growth, new markets and expanded production, initiating some dynamic processes that have reverberated up to the present day. The growing realization that they would have to decolonize their African territories within the foreseeable future also became a powerful incentive during the 1950s for Britain and France, the leading European colonial powers in Africa, to provide these aspiring new states with the technical capacities commensurate with juridical sovereignty. This was the environment in which African countries acquired sovereign status and in which technocratic-conceived plans of development were initiated and implemented.

In political language, the historical rearrangement of Africa's political structures and legal status after 1945 is often subsumed under the terms 'decolonisation' and 'liberation'. Development, rhetorically bound to projects of national liberation, meant obliging communities that were overwhelmingly agrarian to make regular use of minted currencies and exhorting them to become more productive. Systematic attempts were made to encourage individualism and literacy, using techniques of social engineering based on prevailing theories in sociology, political science and the new sub-discipline of development economics. Development required the many societies in Africa – which had been governed throughout their previous existence without reference to political entities that we would today recognize as states, and indeed without the use of writing – to establish large bureaucracies almost from scratch. These measures were applied in a continent that had in most places been quite thinly populated throughout its previous history and only modestly productive in economic terms.

Africa's experience of development bears comparison with the histories of many other parts of the world in the twentieth century. What they have in common are attempts to improve society by the bureaucratic application of policies based on ideologies, theories and techniques conceived in the mode of social science. The American political scientist Zbigniew Brzezinski (1993), who has also served as his country's national security advisor, uses the term 'coercive utopias' to designate these schemes based on purportedly scientific principles to improve the human condition, which were such a marked feature of the twentieth century.

The observation that the development attempted in Africa from the mid-twentieth century onwards could be included in the category of 'coercive utopias' is not a denial that development has brought advantages to many. Development in Africa has brought such benefits as increasing the numbers of children in schools. By improving health care, development policies have brought about the massive increase in population that Africa has witnessed over the last six or seven decades that has been, in the opinion of a leading historian, 'of a scale and speed unique in human history' (Iliffe, 2007, p. 2).

The purpose of these remarks is not to disparage the benefits of development but to consider some of the more damaging consequences of the actual way that development has been applied in Africa, which are less often appreciated. These negative effects include a vast social disruption that is reflected in people's views of the invisible world. Technocratic versions of development and the penetration of capitalism into the heart of African societies became refracted in religious mode. A relationship that Africans traditionally represented in terms of spirits that human beings could contact, and that could even take temporary control of humans in the form of spirit possession, became splintered. It is therefore appropriate to consider in slightly more detail how this occurred and to trace some of its effects.

Religion in sub-Saharan Africa

Religion remains the most important means by which Africans secure access to the invisible world, but today that world is fractured, and communication with it has become difficult.[2] Before wide-scale evangelization and colonization, African practices of communication with an invisible world were woven into daily life and were generally considered an integral part of power as a whole, usually subject to institutional checks and balances. In every African society before colonial times, power was closely associated with authority over the ritual practices that members of a community believed necessary for health, fertility of both land and population, and the reproduction of society – in fact for life itself. It was to a large degree the constant interaction of people with a perceived invisible world that constituted the texture of control over others. From a twenty-first century viewpoint, we can hardly avoid labelling power of this sort as political by nature. Yet few African societies before the late nineteenth century were governed by states even roughly comparable to European ones. Many had no knowledge of writing at all, and in only a few cases was literacy widespread.

Before colonial times, most Africans appear to have thought of the spirit world as, in principle, amoral. That is, the moral stance of an entity in the spirit world depended on the nature of the relationship between a person and a spirit, just as in a relationship between two people. Spiritual power could be invoked for benign purposes (for healing, both personal and social, for example) but also for malign purposes, such as to inflict harm on an enemy. An angry spirit could be pacified and rendered neutral or benign by human action.

By the middle of the nineteenth century, religion was just one aspect of African life that appeared to Europeans as primitive. Europeans, whose commercial and military dominance of the world was unprecedented, had come to believe that they had advanced further in time than any other peoples, and that they were the pioneers of ways of thinking and acting that others would eventually be constrained to follow if they were to survive at all. Africa therefore came to appear as an example of an earlier stage of human organization, a continent of social fossils displaying features of human organization and thought in a stage that Europe had long since left behind.

This way of thinking, underpinned by scientific theories of evolution and racism, informed the administrative arrangements and practices of power of Western origin or inspiration that were imposed on Africans during the age of colonialism. The new conceptual order introduced by European missionaries, colonial administrators and ethnographers extended to the invisible world. Some older techniques, like ancestor cults, were subject to systematic denigration by European missionaries. Some, such as certain forms of blood sacrifice and of initiation, were even outlawed by the colonial authorities. Religious experience as it had previously been known acquired new qualities, new textures and a new moral valence as traditional practices and known spiritual entities were reclassified by European missionaries – many being labelled as 'satanic'. Influenced by missionary ideas concerning the dualism of the invisible world, the African view of the traditional spiritual world became tainted with a suspicion that it was diabolical, backward, and, or, that it should be regarded with suspicion by anyone intent on entering the world of development promised by European technocrats and experts. This is a process that continues to the present day (BBC, 2009). At the same time, as Africans became increasingly drawn within the ambit of colonial power, they adopted new techniques of communication with the invisible world, such as Bible-reading. New forms of religious community, including membership of missionary churches with global aspirations, entailed new forms of sociability and consciousness.

It was not only Africans whose thinking about the invisible world changed during the historical expansion of Western Europe influence to a global level, but Europeans. Europe itself was becoming steadily more secular; religious practice was regarded increasingly as a personal and private matter rather than a political one. The separation of Church and State and even the disenchantment of the world, that is to say move away from a belief in spiritual powers, at least as seen from Europe, are generally acknowledged as important themes of Europe's own history during the age of colonialism. Colonial administrations in Africa, in keeping with convention in their home countries, generally considered religion and politics as two distinct spheres that should properly be subject to institutional and intellectual separation. These same processes did not occur in African societies. What did happen in Africa, however, was that local societies became subject to European ideas in matters of government through the process of colonization and were exposed to a vast range of new intellectual influences. The experience of secular government did not eliminate a widely held belief that the material world was influenced by an invisible world inhabited by spiritual forces.

While ideas concerning the invisible world have changed very greatly in sub-Saharan Africa, even in recent years, one thing at least has remained unchanged – namely, the perception that an invisible world exists. Its sociology and its politics, so to speak, have changed, as has its moral value, but the belief in an invisible world as an actually existing entity has not. 'Essence', in Karl Popper's succinct definition (2002, p. 29), is precisely 'that which remains unchanged during change', and by this yardstick we may say that the perception of an invisible world characterized by its spiritual nature has remained the essence of African religious belief and practice. During the modern age of development that dates from the mid-twentieth century, Africa's invisible world has incorporated new features which are not generally classed as religious at all, but are conventionally considered as economic forces. These include capital itself.

So closely has capital been associated in the Western world with a materialist reading of reality that it may appear incongruous to consider capital as an entity no more visible than an African spirit. One of the characteristics of Western modernity has been the generalization of the belief that reality includes debt and capital, both of which are invisible forces. Yet capital cannot be seen or touched; like a spirit, it becomes visible only when it takes certain forms. Capital is not a material object, but a potential that is deemed to be present in a wide variety of goods and objects. The latent wealth that is constituted by capital can be realized by a series of technical processes, generally considered in terms of law and economics. The most obvious physical form into which capital may be transformed is money. Contrary to popular supposition, capital (which can only be imagined) and money (which can be seen and touched) are not the same thing (De Soto, 2000, p. 43).

Capital, debt and other economic forces or entities have historically become subject to numerical valuation and to manipulation by qualified experts, including professional economists, bankers and accountants. James Buchan (1997), in his brilliant exposition of the history of money, shows how the process of subjecting capital to mathematical control was accompanied by institutional arrangements during the seventeenth and eighteenth centuries, notably in the areas of central banking and insurance. The latter depended on mathematical calculations of the risk attached to various capital-generating activities as well as on the existence of state law. Through the rise of financial institutions and practices of financial mediation controlled by secular organizations and regulated by appeal to secular principles, theologians lost their prior position as the arbiters of 'that redemptive eternity that economists call "the long run"' (ibid., p. 61). It was through the process of applying precise calculation in pursuit of what are assumed to be the regular laws of nature that social sciences, including economics, emerged, and occupied some of the ground previously occupied by religion.

Thus, ideas concerning the invisible world and its relation to the material world can be conceived in various modes, of which religion is only one. In the case of sub-Saharan Africa, incorporation into a world of commerce regulated by capital throughout the last three or four centuries has continued to take place within the idiom of a historically established principle of negotiation with the invisible world. A good illustration of this process concerns the connection between human life, commerce and consumption; the latter imagined as both physical and mystical. It is recorded from places as far apart as Sierra Leone (Shaw, 2002) and Madagascar that the European demand for slaves, that was so important in regulating Africa's entry into global commerce, was interpreted as an appetite for the consumption of human life. People 'had the idea that the Europeans are cannibals', wrote Raombana (1980, p. 245), secretary to the queen of Madagascar, who in the mid-nineteenth century wrote a history of his country in English, 'and that it is the above which compels them to come [...] to buy slaves that they may eat them in their own countries'. Everywhere, the trade in human beings was closely connected with beliefs in the supposed ability to kill or steal a person's soul by mystical means, a power often rendered in English as witchcraft. People accused of witchcraft were often sold into slavery as an alternative to execution (ibid., p. 149).

The close connection formed between sale for money, the consumption of another person, and the theft or killing of the spirit through witchcraft, surely did not exist simultaneously at such distant locations as West Africa and Madagascar as a result of diffusion. It is more convincingly explained as an idea emerging from agrarian societies in which value was expressed largely by reference to fertility and life while the use of currency was quite restricted. In societies where religion, politics and economics did not formerly exist as discrete categories, the forces of capital, the attachment of monetary value to objects and the introduction of minted currencies were all experienced not merely as technical innovations, but in terms of a tectonic change affecting both the invisible world and the material one. Africa's entry into a capitalist world (or, conversely, the entry of capitalism into Africa) has therefore not in itself done anything to cause a separation between the perception of an invisible world that exists, distinct but not separate, conjoined with the visible world. It is by means of interacting with unseen entities, imagined in a variety of forms, that humans shape the world in which they live.

A Wall Street banker may attempt to manipulate capital in his own interest; a religious believer may try to communicate with a spirit for similar reasons. Each perceives the invisible world to exist, and to have recognizable features that are possible to capture for purposes of material benefit. But whereas development experts generally think of this process as a purely secular one, that is most commonly expressed in the language of social science and policy jargon, it has also found its place in the invisible world as it is perceived by many Africans.

European visions of Africa

Belief in an external impulsion to bring progress has been a constant element of European visions of Africa, and has also been present in North American attitudes and government policies. It has informed successive visions of an Africa transformed, passing from the aspiration to free it from the scourge of the slave trade and to bring 'civilization', through technocratic plans and policies to stimulate development and build nations, to contemporary ambitions to bring democracy and human rights to Africa.

The greatest age of development in its technocratic sense was the third quarter of the twentieth century. This was the heyday of modernization, when it appeared to social scientists and policy-makers alike that traditional societies could be fast-forwarded to a modern condition by the application of scientifically verified techniques of social engineering (a phrase apparently invented in the 1950s: Popper, 2002, pp. 37–40). During Africa's golden age of development, it was common for academic observers to ignore religion, assuming implicitly or explicitly that it was due to fade from public importance as countries developed, just as it was thought to have done in Europe.

Since the 1970s, however, a continued rhetorical commitment to development cannot hide the fact that modernization projects have often failed to deliver what was expected of them. Nor can it hide the subversive idea, expressed perhaps first among African intellectuals by novelists such as Chinua Achebe (1966) and Ahmadou Kourouma (1970), and only later by academics such as the political scientist Claude Ake (1996), that Africa's ruling classes were perhaps not primarily concerned to develop their societies and states, but had other priorities that were conveniently masked by an appeal to sublimate politics to the necessities of 'development'. The discourse of development was used by both Africans and their foreign partners to give legitimacy to the edifice of postcolonial governments that were presided over by Africans rather than Europeans, but that continued to make abundant use of the practices, routines and mentalities of their colonial predecessors. These postcolonial states served as a platform for a more ambitious form of political monopoly than anything attempted by the colonizers (Young, 2004). It is in this light that the development project, as actually applied in Africa during the third quarter of the last century, may be included among the many coercive utopias of the last 100 years. Development offered a vision of a future perfection with which whole populations were forced to comply. The accent on human development that is currently championed by the United Nations and many others in the development industry, emphasizing choice and human values over purely economic growth, may be understood as a reaction to the coercion that was inherent in the development vision of the modernizing period.

Not only is it now apparent that the great modernizing vision of development was less than entirely successful, but that it incorporated assumptions about religion that were wrong on several counts. In regard to Europe itself, religion now appears to have played a more important role in Europe's modern history than was perhaps realized 40 or 50 years ago; not least in the extent to which it was co-opted in the earlier twentieth century to produce political religions and sacralized forms of secular power (Gentile, 2006). In Africa, religion has re-emerged as a public force – particularly since the democratization movements of the 1990s – as indeed it has in many other parts of the world. It is therefore striking to re-read the history of Africa in the 1950s and 1960s in the light of our present knowledge, noting that religion never actually disappeared from African societies, even at the time when they were supposedly leaving traditional practices behind. The historical record clearly supports the view that even in the great period of modernization and nation-building, most Africans sought to understand their place in the scheme of things by reference to invisible worlds. The more recent irruption of religion into public space, notably through Pentecostal churches and Islamist movements, does not therefore represent a revival of religion so much as a change in the nature of the relation of religion to the state and to politics. But while the invisible world has never ceased to exist in the perceptions of hundreds of millions of Africans, it has changed radically. It has incorporated elements that in Western thought are separately categorized as religion, politics and economics, and represented as distinct realms of thought and action.

If we are to investigate how changes in the government of material resources may be related to changing ideas concerning the invisible world, it is useful to deploy Michel Foucault's concept of 'governmentality'. Foucault (1994, p. 785) uses this term to designate the particular quality by which power relations may be formed in any given society. Governmentality is situated at the point where technologies of the self – Foucault's term for the methods that every person uses to form and discipline their own personality – meet techniques of subjection and control exercised by others. Governmentality is the specific quality that makes a person or group of people governable in a given historical context. As has been briefly described in the present chapter, religion has played an outstanding role in this regard throughout Africa's history, and continues to do so. Individual self-perceptions and relations within social groups in Africa are formed partly by religious ideas and practice. In the past, this was a key vector of governance in societies that generally had little or no experience of states in anything like the Weberian sense. All this apparatus became formally subjected to colonial systems of power, based on norms that had emerged from the history of the West, including notably the application of written law codes and government by bureaucracy. To this day, these constitute the formal edifice of state power in Africa.

At the same time, a massive change in modes of living brought about by twentieth-century development, including the movement from villages to towns, Western-style education and bureaucratic government, has cut people off from older techniques of accessing the spirit world (cf. Ashforth, 2005, pp. 243–318). As we have seen, in many African societies the invisible world has become more forbidding in the sense that its actions may appear more arbitrary than they once did, and that there are so many competing authorities claiming privileged access to it. Many spirits have ceased to be perceived as entities whose moral nature is malleable or manipulable. The rapid spread of Christianity and Islam in the twentieth century challenged this perception since both are dualistic religions, with God as the supreme good and Satan the prince of evil. Consequently, many traditional African spirits have become perceived as evil by nature. The evident failings of modern societies and governments are often attributed to the existence of cosmological forces that politicians, being powerful people, are called upon to master. Since politicians often fail to preside over prosperity and harmony, it is often assumed that their motives are malign. A feeling of powerlessness is widespread. Not least, the growth of bureaucratic states has challenged people's ability to channel spiritual forces because these states, being based upon secular principles, have no competence to deal with the spirit world.

To avoid any misunderstanding, it should be said that people who perceive prosperity partly in terms of a relationship with the invisible world are not necessarily ignorant of basic economics. A religious mode of apprehending reality constitutes an epistemology that is simultaneously traditional and modern, capable of updating and renewing itself as times change. Above all, African epistemologies, couched in a religious idiom, offer a theory of causation with which secular views are unable to compete, since a secular world-view is unable to offer a satisfactory explanation for why some people are fortunate, while others are not.

Thinking with religion

'Many African leaders believe that the international economy is still rigged so that Africans will never prosper', so we are told by the US National Intelligence Council (2005, p. 8). The material and political reasons for such a feeling of powerlessness are widely understood, as analyzed by a vast literature in social science. Less widely appreciated is the degree to which Africa's people feel themselves to lack power because they are not able to shape the invisible world to their own requirements in ways that once seemed possible, by communicating with spiritual entities. Detailed investigations of this state of affairs, for example by the brilliant Congolese sociologist Joseph Tonda (2005), use the paradigm of the invisible world to explain both the outrageous corruption of some governments as well as some of Africa's more striking social ideas and attitudes (such as the cult of consumption that is a feature of life in West-Central Africa) by reference to changing attitudes towards the ancestors, debt, capital and other such invisible entities.

Few anthropologists or sociologists these days subscribe to the modernization theory, which is thoroughly discredited in academies of learning. Yet, according to Achille Mbembe (2001, p. 7), some of the key disciplines that contribute to debates on development and on the formation of policy by powerful bureaucracies actually continue to hold many of the basic assumptions underlying modernization. The literature of political science and development economics, Mbembe claims, continues to be 'almost entirely, in total thrall' to the teleology represented by 'theories of social evolutionism and ideologies of development and modernization'. He asserts that 'these disciplines have undermined the very possibility of understanding African economic and political facts' (ibid.).

Thus, there is a challenge both to rethink the nature of development and the disciplines through which we approach it (cf. Ellis & ter Haar, 2007). This challenge is rather more testing than a standard academic appeal to cast off an old paradigm and replace it with a new one, since it means understanding societies that conceive of prosperity in relation to an invisible world that has been systematically excluded as a serious source of knowledge from key branches of the social sciences. The days are long gone when we could suppose that various processes that have played themselves out in the history of Europe and North America, such as the separation of politics and religion, and Church and State, must occur in all societies as they progress through time. If we seek to know how Africa might develop in the twenty-first century, it is necessary to include religion.

Notes

1. The text of the speech is available at http://www.bartleby.com/124/pres53.html [accessed 16 April 2010].
2. Gerrie ter Haar and I have elsewhere defined religion in Africa today as 'a belief in the existence of an invisible world, distinct but not separate from the visible one, that is home to spiritual beings with effective powers over the material world' (Ellis & ter Haar, 2004, p. 14).

References

Achebe, A. C. (1966), *A Man of the People*, London: Heinemann.

Ake, C. (1996), *Democracy and Development in Africa*, Washington, DC: The Brookings Institution.

Andrew, C. & Mitrohkin, V. (2005), *The Mitrohkin Archive II: The KGB and the World*, London: Allen Lane.

Asad, T. (1993), *Genealogies of Religion: Discipline and Reasons of Power in Christianity and Islam*, Baltimore, MD: Johns Hopkins University Press.

Ashforth, A. (2005), *Witchcraft, Violence, and Democracy in South Africa*, Chicago: University of Chicago Press.

Bayly, C. A. (2004), *The Birth of the Modern World, 1780–1914*, Oxford: Blackwell.

BBC News (2009), 'Witchcraft in West Africa', 30 August, http://news.bbc.co.uk/2/hi/africa/8229203.stm [accessed 3 September 2009].

Brzezinski, Z. (1993), *Out of Control: Global Turmoil on the Eve of the Twenty-First Century*, New York: Charles Scribner's Sons.

Buchan, J. (1997), *Frozen Desire: The Meaning of Money*, New York: Farrar, Straus and Giroux.

Cohn, N. (1957), *The Pursuit of the Millennium*, Oxford: Oxford University Press.

Cooper, F. (2002), *Africa Since 1940: The Past of the Present*, Cambridge: Cambridge University Press.

De Soto, H. (2000), *The Mystery of Capital: Why Capitalism Triumphs in the West, and Fails Everywhere Else*, New York: Basic Books.

Ellis, S. & ter Haar, G. (2004), *Worlds of Power: Religious Ideas and Political Practice in Africa*, London: C. Hurst & Co.

Ellis, S. & ter Haar, G. (2007), 'Religion and Politics: Taking African Epistemologies Seriously' *Journal of Modern African Studies*, 45 (3), pp. 385–401.

Ellis, W. (1838), *History of Madagascar*, vol. 2, London and Paris: Fisher, Son, & Co.

Foucault, M. (1994), *Dits et écrits*, vol. 4, Paris: Gallimard.

Gentile, E. (2006), *Politics as Religion* (trans. G. Staunton), Princeton, NJ: Princeton University Press.

Iliffe, J. (2007), *Africans: The History of a Continent*, 2nd edn, Cambridge: Cambridge University Press.

Kourouma, A. (1970), *Les Soleils des indépendances*, Paris: Editions du Seuil.

Lal, D. (1998), *Unintended Consequences: The Impact of Factor Endowments, Culture and Politics on Long-Run Economic Performance*, Cambridge, MA: Massachusetts Institute of Technology.

Leys, C. (1996), *The Rise and Fall of Development Theory*, London: James Currey.

Masuzawa, T. (2005), *The Invention of World Religions: Or, How European Universalism was Preserved in the Language of Pluralism*, Chicago: University of Chicago Press.

Masuzawa, T. (2007), 'Theory Without Method: Situating a Discourse Analysis on Religion', in G. ter Haar and Y. Tsuruoka (eds), *Religion and Society: An Agenda for the Twenty First Century*, Leiden: Brill.

Mbembe, A. (2001), *On the Postcolony*, Berkeley and Los Angeles: University of California Press.

National Intelligence Council (2005), 'Mapping Sub-Saharan Africa's Future', Conference Report: http://www.dni.gov/nic/confreports_africa_future.html [accessed 28 December 2008].

Pagden, A. (1986), *The Fall of Natural Man: The American Indian and the Origins of Comparative Ethnology*, Cambridge: Cambridge University Press.

Popper, K. (2002) [1957], *The Poverty of Historicism*, London and New York: Routledge.

Raombana (1980), *Histoires*, vol. 1, Fianarantosoa (trans. S. Ayache), Madagascar: Librairie Ambozontany.

Rist, G. (2002) [1996], *The History of Development: From Western Origins to Global Faith*, London and New York: Zed Books.

Shaw, R. (2002), *Memories of the Slave Trade: Ritual and the Historical Imagination in Sierra Leone*, Chicago: University of Chicago Press.

Tonda, J. (2005), *Le Souverain moderne: le corps du pouvoir en Afrique centrale (Congo, Gabon)*, Paris: Karthala.

UNDP (2009), 'The Human Development Concept', http://hdr.undp.org/en/humandev/ [accessed 2 September 2009].

Young, C. (2004) 'The End of the Post-Colonial State in Africa? Reflections on Changing African Political Dynamics' *African Affairs*, 103 (410), pp. 23–49.

2
The Mbuliuli Principle: What is in a Name?

Gerrie ter Haar

Religion and development: A controversial debate

In the wake of a renewed general interest in the role of religion in public life, development agents have begun to reflect on the relation between religious thought and development practice. This is distinct from the substantial literature that has been produced on the role of institutions that channel religious ideas, known as faith-based organizations or FBOs (e.g. Marshall & Keough, 2004; Tyndale, 2006; Marshall & Van Saanen, 2007). Well-known FBOs include international organizations with a Christian background, such as Caritas or World Vision, Tearfund or Christian Aid to mention only a few examples widely known in Europe. There are also many other faith-based organizations that have been, inspired by different types of faith. These include major organizations such as Islamic Relief or the Aga Khan Foundation, but also numerous ones of Hindu and Buddhist extraction that are operating mainly in parts of the world where a majority of the population adhere to these religious traditions. In fact, it appears that, considered from a world perspective, religion provides an important inspiration for many people to engage in development programmes (e.g. http://berkleycenter.georgetown.edu; http://www.religon-and-development.nl). In recent years, this fact alone has caused policy-makers of all sorts, including governmental as well non-governmental agents, to pay attention to the factor of religion in designing their policies.

However, though it is quite common today for development agents to include religious organizations in their work, there is still a great reluctance to investigate the type of ideology on which these institutions are based. This is largely due to a fear on the part of secular institutions and of researchers educated in a secular tradition of being drawn into a sphere of life which – in the West – is seen as not belonging to the public realm. Yet there is a growing awareness that the current state of affairs is unsatisfactory in this regard. Not only has religion visibly reoccupied public space in all parts of the world, but the very efficacy of Western development efforts is at stake. In the Netherlands, this has led to a unique initiative in the establishment of a Knowledge Centre Religion and Development (KCRD), in which Dutch development organizations in collaboration with some academic counterparts embarked on a process of re-examination of their ties with development organizations in the south. The Knowledge Centre 'seeks to put into practice the vision that religion is an important factor for sustainable development, international co-operation and civil society building' (http://www.religion-and-development.nl). To fulfil that aim, it collaborates with various actors in the field of development, notably bringing together theoreticians and practitioners, and acting as a broker in the field. This activity started with a series of consultations culminating in two large conferences in 2005 and 2007 respectively, in which some 100 participants from all parts of the world and from different religious backgrounds discussed the need to change the nature and direction of development co-operation (Various authors, 2005, 2007).[1] This reorientation was instigated by an awareness of the shortcomings in framing development as an exclusively secular project. According to the Knowledge Centre, lack of knowledge about the role of religion in people's lives can have far-reaching consequences for the practice of development co-operation, the improvement of human rights, and for building civil society.

A major obstacle to investigating the role of religion in development is a widespread misunderstanding about the meaning of religion. It appears that in the so-called developing world 'religion' needs to be understood in the broad sense of a belief in an invisible world inhabited by spiritual entities or forces that have a bearing on people's material lives.[2] In other words, for most people in the world 'religion' refers to a way of viewing the world rather than to some organized system of belief to which they subscribe. The invisible world in which they believe is distinct but not separate from the visible one; it is a world with which they can interact and communicate regularly for their own benefit, both materially and spiritually. For them, this is a world that contains spiritual power, a type of power which they can share in fully – and in that sense it is rather unlike political power, which tends to be more exclusive. Spiritual empowerment, as we will come to see, is an effective strategy from a believer's perspective, since it opens up avenues to achieve what is often referred to as the 'good life', which is what development is supposed to be all about.

In this context, we may therefore note that religion, in the sense here defined, is a driving force in many communities. It can be – and often is – employed for the common good.[3] The resilience of religion in the southern hemisphere has come as a great surprise to many observers, academic and other, who had considered the 'disenchantment' of the world as an inevitable outcome of the modernization process. The fact that reality has proved to be other than predicted suggests a need to re-think some fundamental approaches to questions about how societies evolve. Perhaps the greatest problem in this regard is that social science has, from its inception, been based largely on precepts drawn from an idealized reading of European history, on the assumption that European societies have been the most advanced in historical development and that they therefore offer pioneering models for societies elsewhere. If we accept that this is no longer a reasonable assumption, then it requires us to reconsider many underlying assumptions concerning development, and the institutional arrangements that are based on them. 'Disenchantment' came to be considered by sociologists as an inevitable accompaniment of the rise of modern states and modern economies because European societies were thought to have become 'disenchanted' over time, and therefore classical theories of development have paid no attention to religion. To development experts, religion seemed irrelevant to the processes they were analyzing, except as an obstacle to modernization. However, it is now clear that these assumptions do not hold true for many other parts of the world, including Africa, where religion has continued to play a central role in people's lives, both privately and publicly. We may say that the continuing role of religion in public space in Africa represents a form of historical continuity with the continent's deeper past.

Particularly through colonialism, it appears, Europeans transferred their own historical experiences to Africa – and other parts of the world, for that matter – by imposing the separation of the realms of the visible and the invisible, notably in the form of the separation of religious and political power. In many cases, this modernist vision of politics has not brought the welfare and prosperity expected at the time of independence. Hence, many people in the non-Western world, including in Africa, are now reconnecting with their own history and relying on their own traditional resources by seeking ways of giving religion a legitimate place in their societies. Although they may continue intellectually and practically to distinguish between the spheres of the religious and the secular, or between the realm of the Church on the one hand and of the State on the other, they may no longer wish to separate the two on grounds that find their historical justification in the specific history of the West (ter Haar, 2007).

The fact that the vast majority of people in the world are religious in some shape or form is itself sufficient reason for taking the religious dimension very seriously in the development debate. But there are other reasons too for considering the role of religion in development. Religion, in whatever form it may manifest itself, and irrespective of the analyst's personal likes and dislikes, constitutes a social and a political reality. For many people in the world religion is a powerful motivation to act in the ways they do. It provides them with the moral guidance and inspiration to try and change their lives for the better. Many examples could be cited from recent history from all parts of the world, including Poland, South Africa and the Philippines, to mention only three countries in which religious inspiration contributed significantly to major social and political changes. For most people in the world religion is part of their social fabric and it provides one of the main ways in which they choose to organize themselves. Many people voluntarily associate themselves with religious networks, which they use for a variety of purposes – social, political and economic – that go beyond any strictly religious aspect. All this is extremely relevant to the development debate.

But, in the end, the most important reason to pay serious attention to the religious dimension of people's lives, in my view, lies in the need to maximize resources for development.

Mobilizing spiritual resources

Development is thus not to be equated only with economic development. This is an insight that – even if reluctantly – most development actors have become aware of in recent years, whether at the governmental or intergovernmental, national or international level. A World Bank initiative, starting with the World Faiths Development Dialogue, is an important marker in this regard.[4] Shifts in perception have given rise to the idea that it is helpful to identify the positive potential of religion for development and to consider ways of mobilizing this asset. The resources available for development include not only material and intellectual resources, but may also include religious or spiritual ones. The dominance of the economic paradigm in development co-operation is probably the principal reason that religion has not yet been allowed to use its full potential for human development, other than through service delivery. Its inherent potential – also referred to as spiritual capital – has hardly been mobilized. The mobilization of religious resources was identified by participants of the conferences in the Netherlands, referred to above, as a crucial dimension of what was dubbed 'integral development'. The latter is a notion in need of further explication, but the spiritual dimension is no doubt an important element in it.

In their book *Spiritual Capital: Wealth We Can Live By* (2004) Zohar and Marshall identify spiritual intelligence (SQ) as a particular form of human intelligence, different from rational intelligence (IQ) and emotional intelligence (EQ). They argue that spiritual capital forms the underlying base of any other kind of capital, including capital of a material sort. Spiritual intelligence, in their view, is what provides a sense of meaning and values and fundamental purpose on which to build spiritual capital, which can generate wealth. 'It is only when our notion of capitalism includes spiritual capital's wealth', they argue, 'that we can have sustainable capitalism and a sustainable society' (ibid., p. 4). People, organizations and cultures that have spiritual capital, they believe, on both subjective and scientific grounds, will be more sustainable than those without it because they will have developed a broader range of qualities, in particular addressing concerns about what it means to be human. To mobilize this potential, a more comprehensive approach to development is needed, taking the indivisibility of the human person as a point of departure.

If we take this approach seriously, it will require new paradigms in development, as was stressed in the two major conferences, mentioned above, that were held in the Netherlands in 2005 and 2007. It means that the concept of development itself is in urgent need of redefining, in such a way as to incorporate both the material and the immaterial dimensions of human life. According to religious believers, the concept of change, which is intrinsic to development, should be linked to the idea of personal transformation, acknowledging the connection between the individual and social dimensions of change.

Spiritual capital, in other words, may be seen as part of social capital. Social capital is an umbrella term that can be defined in a variety of ways. Most definitions contain references to the degree in which communities are able to work together for the common good through such mechanisms as networks, shared trust, norms and values. Some refer to the value of social networks that people can draw on to solve problems, in which case social capital is associated with trust, reciprocity, information and co-operation. Definitions of social capital may also refer more specifically to the attitude, spirit and willingness of people to engage in collective, civic activities, such as the skills and infrastructure that aid in social progress. If we compare these different perspectives with the World Bank definition of social capital as 'the norms and social relations embedded in social structures that enable people to coordinate action to achieve desired goals' (http://www.worldbank.org/), we can see that important recurring notions in the various approaches are norms and values, trust and relationality (cf. http://www.spiritual capitalresearchprogram.com). Inasmuch as the invisible world is an integral part of the world, as people know it, spiritual capital provides an additional social resource for those who actively engage with the invisible powers they believe to exist. In that sense, so-called FBOs may be considered to have access to potential resources – spiritual or religious resources – that secular NGOs normally lack.

An African view on development: Archbishop Milingo's approach

Many of the insights that have emerged in recent years regarding the role of religion in development and the need to be attentive to the spiritual dimension of human life are less innovative than is sometimes apparent. It may be said that, generally speaking, development theorists and policy-makers are only now beginning to see an aspect of the problems facing them. This was previously obscured by a secular worldview that identified human development with material progress, and assumed that religion would decline in importance in the face of secularization. As long ago as the mid-1970s, the Zambian archbishop Emmanuel Milingo, who was then Archbishop of Lusaka, was expressing trenchant criticisms, verbally and in writing, of the development approach of many Western development agents. His views were never given serious attention, no doubt in part because they came from a Catholic prelate at a time when mainstream development thought was primarily of a secular nature. Furthermore, Milingo became known in the West almost exclusively as a faith healer and exorcist (ter Haar, 1992). The exclusive focus on this particular aspect of his work has turned Milingo into an exotic and controversial figure in this particular part of the world. In 1982 he was called to Rome by the Vatican and subsequently made to resign as Archbishop of Lusaka. Following this hearing the Church did not allow him to return to Zambia, and Milingo has been living in the West ever since.[5]

Most importantly, Milingo's reputation in the West has obscured the fact that he has been, and continues to be, deeply engaged with development issues. This is clear from his biography which outlines this development work that includes, among other things, founding the Zambia Helpers' Society, a voluntary organization established in 1966 to cater for the medical needs of people in the shanty towns of Lusaka, still in existence today.[6] He was also the initiator of the Fast Learning Project, set up that same year with a view to teaching people how to read and write by the year 2000 (Milingo, 1966), and the writer of *Amake Joni* ('Joni's Mother'), intended to prepare young adults for parenthood while preserving their family tradition (Milingo, n.d.). The book was later approved by the Zambian government for use in schools. These are all initiatives that go back to the 1960 and 1970s.

Milingo's development initiatives were integrated into criticisms of African governments that he made both implicitly and explicitly, at a time when African nationalism was largely exempt from criticism in the world of development (Milingo, 1971). Archbishop Milingo consistently denounced African governments for their greed for money and political power, comparing them even to 'witches, whose ritual includes evil as a means to carry on their profession' (Milingo, 1994, pp. 26–7). He blames African political elites for exploiting their own people and keeping them poor. He has long regarded this state of affairs as a reason why things have not improved much since independence and an explanation for what he terms the deplorable state of affairs on the continent. Some three decades ago, Milingo described the transition from colonial rule to independent government as one of coming out of the mouth of a lion and going into the leopard's mouth. 'The poor man is not safe, he dies all the same, because both these animals are carnivorous' (ibid., p. 4).

Perhaps it is because of Western academic sensitivity to the perceived spiritualization of politics in Africa that Milingo's political critique has gone largely unnoticed. Much of his critique from 30 or more years ago resonates with what appears to have become mainstream thinking in the twenty-first century. As early as in 1976, Milingo described in a pamphlet how the word 'development' had become an item of jargon among social welfare promoters, with a negative effect on developing communities. His thinking about development displays the holistic approach which is so evident in African religious worldviews generally, particularly in Africans' experiences of illness and disease, which were the crucial factor in Archbishop Milingo's healing ministry (ter Haar, 1992). Generally speaking, healing in Africa is an all-embracing concept that is not restricted to physical healing, but also addresses the mental and spiritual dimensions. Healing implies the full restoration of human dignity, which is offended by oppression of every kind. Illness is perceived as any sort of stumbling-block on the road to human fulfilment. In order to heal a person, offending obstacles need to be removed. From this point of view, poverty is also seen in terms of a type of illness that requires healing, and this is how it is often experienced. Milingo – and many other African clergy for that matter – frequently refer to the 'spirit of poverty', implying a form of evil from which Africans can liberate themselves through spiritual action or communication with a spirit world.

As part of his comprehensive approach to life Milingo refuses to equate human development with the material progress of a nation. 'The material heaping of goods into the hands of the people', Milingo wrote in the 1970s, 'just makes them scramblers and consumers. As a whole they become irresponsible people, expectant of every good thing at the receiving end' (Milingo, 1994, p. 3). In his view, this is largely because 'development' is seen as synonymous with 'progress'. He makes a distinction between these two terms. While he considers progress as purely material advancement and therefore in relation to what he terms 'external' (i.e. consisting in outward acts) human achievement only, the term development should be understood in his view as a gradual evolution or completion, hence the result of a process. External progress, he argues, does not guarantee the fullness of humanity that true development would bring. It leads him to the following definition of development: 'The freeing of human potentialities to be channelled to the profitable use for the individual himself as well as for the community in which he is living' (ibid., p. 4). This process must lead individuals to a change of attitude towards life, and help them to realize what they have (their potentialities), and what they can do for themselves (create self-confidence). Development, thus, should aim at educating people to be aware of what they are and what they can do.

Self-confidence and self-reliance, as we will come to see, are key concepts in Milingo's thought about development. These are notions that explain why Milingo has been consistently critical of the development theories of Western agencies and aid workers who, in his view, have not really taken into consideration the role of ordinary people in the communities where they live. The experts arrive, he claims, with ready-made answers, carrying out projects which may be well-intentioned but that basically disturb people's lives, after which they leave, complaining about the lack of gratitude on the part of African people. Or, as he put it in his own metaphorical language: 'If you are told to cover a dead body with a cloth you may take any colour. But if the dead person were alive, you should first have asked him as to whether he accepts or not this or that colour' (ibid., p. 34). Development projects, Milingo has always believed, should begin with the people concerned. This is a point of departure that seems to be increasingly accepted today by development agents, as is reflected in the title of the present volume, which focuses on 'development from below'.

Milingo's views of development have changed little since he started to reflect systematically on the matter in the 1970s. More than 30 years ago he was describing what nowadays all and sundry seem to agree upon, namely that Africa has been turned into a continent of beggars and made dependent to an alarming degree on Western development aid. To his mind, this is largely due to the dominant Western paradigm that defines development purely in terms of material progress and has turned Africans into consumers. Milingo insists that human development – a term popularized since the publication of the first United Nations Human Development Report in 1990 – cannot simply be equated with the material progress of a nation, and the latter should not be seen as a criterion of civilization. 'Wearing a suit', as he puts it, 'does not necessarily exclude one from the membership of bandits' (ibid., p. 4).

The Mbuliuli principle

In recent years, while he was still living in Rome, Archbishop Milingo resumed his reflections on the issue of development in Africa, and he has attempted to put his ideas into practice in his home country, Zambia. His more recent ideas provide an interesting example of the way in which development theory can find a point of departure in the daily experiences of the people concerned, in this case ordinary Zambians. Milingo has dubbed his theory the Mbuliuli Economic Principle, the details of which he has elaborated on in three booklets which have been organized in the form of short instruction books or user guides (Milingo, 2004–5). The name *mbuliuli* refers to the maize that is the staple diet for most Zambians, more particularly 'the maize which bursts and takes a new form (popcorn)' (ibid., *Phase One*, p. 22). It is the type of maize (popcorn) that bears the name of its inherent potential: it is what it will be when put in a frying pan. In other words, *mbuliuli* is a grain that carries the characteristic of inner change or transformation. In Milingo's creative mind the *mbuliuli* principle is turned into the leading principle of a home-made economic theory based on local knowledge – although he realises full well that his theory is not going to make what he calls 'a soft landing' (ibid., p. 3) in professional economic circles. It is a very practical theory, based on the fundamental belief that economic prosperity is linked to inner growth, or, that economic expansion, or external expansion, goes hand in hand with 'expansion from within' (ibid., p. 2).

Self-reliance and sufficiency are the core elements in the *mbuliuli* approach. Applied to the material side of life, or economics, *mbuliuli* refers to expansion from within an individual enterprise, which may take various new forms (ibid., p. 5). The profits will be ploughed back into a local community to increase its capital. In this regard, Milingo fills conventional economic terms with new meanings that are both connected with the cultural knowledge of ordinary Zambians (ancestral wisdom) and do justice to their religious worldview ('traditional' and Christian in this case), as well as being presented in their own Zambian languages. The *mbuliuli* principle is equated with *mayipezi*, a Nguni word that is used to describe the situation before or after the rains. In the same way as the word *mayipezi* refers both to the warning and the hope of the rains, so does the *mbuliuli* economic principle contain both a warning and a hope of better times. The working of the *mbuliuli* economic principle is compared to the work of a cultivator during the rainy season (ibid., *Phase Two*, p. 19). In his exposition of how local communities can take care of themselves, Milingo introduces new terms, derived from African languages and explained in English, such as 'fluid money' and 'capital donation', and he refers to the local community that owns the means of production as a 'Moral Entity'. The head of a so-called Mbuliuli Enterprise is referred to as the 'care-taker' of the business, that is characterized by joint ownership.[7] Several such Mbuliuli Enterprises are now in existence in Zambia.

All this may not be easy to follow for those versed in more conventional development jargon. But in the context of an approach that favours 'development from below', it becomes important to understand the language in which individuals and communities express themselves, including when ideas are couched in religious terms, and the self-understanding on which such language is based. In this case, the basic principles are growth or expansion from within, or 'potentiality'. In Milingo's view it is this, and not some external force, that will move Africa's economies forward and will bring prosperity and abundance. Economic productivity must be rooted in local thought and practice, reflecting the cultural values of the communities concerned. Zambia, he argues, has never had its own economic foundation, but has been 'spoon-fed' through economic imitation, both during colonization and after independence. Since then the country has been economically hit by privatization, both of its mines and of other sectors of the economy, and is now being overrun by economic globalization. The end result is that the nation's wealth has left the country. 'We Zambians have remained like young pigeons in the nest opening out mouths, waiting for our mother to drop in something to eat. And nothing is coming in' (ibid., *Phase One*, p. 2). Zambia's development depends on the answer to the question: 'Where are we and where are we going?' (ibid., p. 3). The *mbuliuli* principle is intended as the underlying idea on which to build a stable homemade economic theory, taking into account insights handed down over the centuries, notably in the form of ancestral wisdom.

To illustrate these points, it is useful to quote an example of the *mbuliuli* economic principle in action, related in Milingo's own words:

> The Twende Bus Company was going to ruins.It was going to die a gentle death, but all the same a sad and tragic death. Through the **Mbuliuli Economic Principle**, keeping the name Twende, but attaching it to 'the Lyods Carriers', Twende began to raise it head up, without shame. Today Twende is speaking of possible travel adventures to Mongu, with only two buses. It is because it has the support of Lyods Carriers.
>
> It is not just a make belief that Twende can be supported by Lyods Carriers. Before long Lyods Carriers will have a float [*sic*] of four big lorries, the maximum carriers in Zambia. The income from lorries is fixed, different from that from buses. It is a much more stable income. There is however a need of common economic strategy. With the income for instance from the lorries, several grinding mills can be bought. The grinding-mill in a month will triple the income of a lorry. It is at this time when stealing comes in. Because the money from grinding-mills is often cash. And if the receiver of money is not different from the operator of the grinding-mill the money will slowly diminish, and finally will not be forthcoming.
>
> (ibid., p. 6)

What is of interest here is not so much the practical details that characterize the initiative, but the ideas that underpin the introduction of the Mbuliuli Enterprises. These are built on two pillars: religion and culture. On the one hand, Milingo's ideas about development are clearly inspired by Catholic social teaching with, at times, a radical slant that recalls secular social theories of the late twentieth century (cf. e.g. *ujamaa* in Tanzania); on the other hand they are inspired by his firm conviction that development practice must be embedded in the way of life of the people concerned. The latter must also be considered in view of Milingo's lifelong struggle to help Africans overcome the historical burden of slavery, colonial domination, economic exploitation and racial discrimination, which have left many of them with a serious inferiority complex. It is on these grounds that he strongly objects to common references to Africa as an underdeveloped continent, without further explanation of its present status. On many occasions, he has voiced his views on the way in which 'Africa has been systematically and economically milked' (Milingo, 1994, p. 16) since the arrival of the first Europeans, the Portuguese, in the fifteenth century. Restoring Africans to their human dignity is the main driving force behind Milingo's work and thought in all respects. Restoration of their dignity is needed, in his view, before they can successfully assume their tasks in society and contribute to the development of humankind.

Milingo's elaboration of the *mbuliuli* approach to development is conceived as a way to install self-confidence rather than to rely on external forces to bring about change. It is a logical sequel to the controversial healing ministry for which Milingo has become best known. To him, healing simply implies taking away any disturbance that prevents people from being fully human, thereby restoring them to their original dignity. This way of thinking motivates his thought and action concerning development. The *mbuliuli* approach is an expression of the belief that for lasting effect, or sustainability, one must start at the individual level, with confident and self-reliant people, unharmed by negative forces from outside, spiritual or material. This is the condition, he believes, that will bring a new business spirit to Africa and help it to arrive at self-reliance rather than being dependent on the exploitative aims of foreign investors that leave the continent poor (Milingo, 2004–5, *Phase One*, p. 33). It is an approach that recognizes that development is intrinsically connected to human dignity, both of a country and of its citizens.

Religion and development: an alternative agenda

In the present chapter, Archbishop Milingo's *mbuliuli* principle has been used as a metaphor for an alternative approach to development that is driven by local understanding. By way of conclusion I will place

Milingo's ideas in the wider context of current debates on religion and development, and highlight some points of consonance and conformity. I will mention six such points which, I suggest, are also important points of analysis in future debates on religion and development.

1) The most important aspect of the metaphor used by Milingo is the aspect of **transformation**, represented by the inherent nature of *mbuliuli*. Not only does the name refer to the staple diet of many Zambians but, more importantly in the context of our argument, the essential characteristic of *mbuliuli* is its potential to become something else, that is, to transform itself. Transformation appears to be a key concept in the current development debate, implying the need and the capacity for radical change at various levels, personal and social, in that order. The awareness that such drastic change is actually possible, and the envisaging of a totally new life, is one of the most helpful ideas in existing religious ideologies in relation to development.

2) Central to the debate on religion and development is the issue of **self-realisation** through the use of the full spectrum of human potentialities and the need for people not only to take responsibility for their own lives, but also to be allowed to do so. According to the United Nations Development Programme (UNDP), human development 'is about creating an environment in which people can develop their full potential and lead productive, creative lives in accord with their needs and interests' (http://hdr.undp.org.hd/). This resonates with Milingo's definition of development as 'man's contribution to self-uplifting to a better living condition and to the realisation of his own potentialities to make him a full person' (Milingo, 1976/1994, p. 41). In both cases the meaning of development is taken to refer to people's resources beyond any purely material and technocratic aspect.

3) This has a bearing on the fashionable notion of **agency**, something that is much invoked but not necessarily observed in the business of development. It deserves more prominence in development practice, if recipients of aid are to be not only objects but also subjects of development co-operation. This also implies a greater attention for local initiatives in development co-operation, with implications for agenda-setting. Milingo's *mbuliulu* approach is an illuminating example. Such an approach, taking the notion of agency seriously, builds on what people have and not on what they lack. It helps them to realise their potential.

4) Development co-operation suggests a relationship. One way of looking into the spiritual potential of religion, it has often been suggested, is as a way of relating to the 'other'. '**Relationality**' was identified in the international conferences in the Netherlands as a key concept for theoretical and practical exploration in the relation between development partners in the rich and poor parts of the world. 'Inreach' is part of that process and refers to the need for self-reflexivity on both sides: Western donors and the recipients of aid. This was captured in the phrase: 'No outreach without inreach' (*Transforming Development*, 2007). The latter was deemed particularly important for faith-based donor organisations in the West, which need to reflect on their faith-based identity and the implications this entails. Milingo has made a comparable point in stating that '[M]aterial progress does not always prove that the one who gives it has a pure intention. It is not a proof either that the relations between the giver and the receiver are harmonious' (Milingo, 1994, p. 35). The inherent tension in the relationship between Europe and Africa that emerges from its history still needs to be resolved.

5) The issue of **social trust** is deemed vital in the religion and development debate. This is argued by Milingo when he emphasises the need first to establish human friendship, from which mutual trust will then follow (i1994, p. 37). The concept of 'human development' as proposed by the United Nations broadens the concept beyond its economic meaning to include, typically, a wide range of human relationships. From a religious perspective, the economic paradigm is detrimental to development co-operation if it is not balanced by the broader idea of human development. By incorporating the aspect of human development, the concept of economy will be expanded to include what has been called an 'economy of caring and sharing' (*Religion: A Source for Human Rights and Development Cooperation*, 2005, pp. 17, 34, 37). This, we saw, is the basis of Milingo's Mbuliuli Economic Principle.

6) For Archbishop Milingo, the key issue in the development debate centres on the question of **human dignity and freedom**. This is highly relevant given the predominance of the so-called rights-based approach in development. It has been argued by some that today's rights-based development discourse needs to be critically interrogated and compared with other approaches in order to see where it originates, who is articulating it, what are the differences in the versions used by different development agents and what are their shortcomings, and what implications these have for the practice and politics of development

(Nyamu-Musembi & Cornwall, 2004, p. 1). A religious or spiritual approach to development is one such form of critical interrogation and engagement with secular approaches – not to replace these but to complement them. It is worth noting in this regard, the faint echo of Amartya Sen's ideas concerning development as freedom (Sen, 1999).

More points of consonance between the ideas of Archbishop Milingo and the general debate on religion and development could be mentioned, such as the need for contextualization in development co-operation, the emphasis on a renewal of values, and the importance of what terminology is used. Words are important since the language that is used is indicative of the way in which outsiders shape other people's reality or, conversely, deny other people their own views of reality. For many people in the world, their reality includes the invisible world, and, to this extent, their worldview may be considered 'religious'.

Effective development, it may be said in conclusion, requires a mobilization of a full range of human resources. It should start from people's own worldviews, which in many cases are religious. This does not mean an exclusively spiritual approach, but an effective combination of the secular and the religious. As an Indian colleague has stated: 'Those who work for materialist purposes only live in darkness, those who work for spiritual purposes only live in greater darkness.'[8]

Notes

1. A report of the 2005 conference is available at http://www.icco.nl/; a report of the 2007 conference is available at the KCRD website http://www.religion-and-development.nl/
2. This broad definition rests upon the work of E. B. Tylor, who defined religion succinctly as 'the belief in Spiritual Beings'.
3. This does not imply that religion cannot equally be employed for purposes that are harmful to individuals and communities.
4. It was former World Bank President James Wolfensohn who initiated a series of dialogues between the Bank and religious leaders. Together with former Archbishop of Canterbury, George Carey, he convened the first Faith and Development Dialogue held in London in February 1998.
5. He now lives in Washington as a married man, associated with the church of Reverend Sun Myung Moon.
6. In recent years, the Zambia Helpers' Society has among others built a hospital in the outskirts of Lusaka.
7. This is comparable to the idea underlying to rotating credit unions in Africa, such as *tontines* or *susu*.
8. Chander Khanna during the conference 'Transforming Development: Exploring Approaches to Development from Religious Perspectives', held in

Soesterberg, The Netherlands, in October 2007. See http://www.religion-and-development.nl/

References

Marshall, K. & Keough, L. (2004), *Mind, Heart and Soul in the Fight against Poverty*, Washington, DC: The World Bank.

Marshall, K. & Van Saanen, M. (2007), *Development and Faith: Where Mind, Heart and Soul Work Together*, Washington, DC: The World Bank.

Milingo, E. (n.d.), *Amake Joni* [Privately published before 1970].

Milingo, E. (1966), 'Fast Learning: How to Read and Write in the Year 2000' [Privately circulated].

Milingo, E. (1971), 'Patronado and Apartheid: Easter Message', Published in *Herder Correspondence*, under the title 'A voice across the border'.

Milingo, E. (1994), *Development: An African View*, Broadford, Victoria: Scripture Keys Ministries Australia [originally published in 1976, mimeograph].

Milingo, E. (2004–5), *The Mbuliuli Economic Principle. Phase One; Phase Two* (n.d.); *Phase Three* (2005). Zagarolo, 2004–5 [Privately circulated].

Nyamu-Musembi, C. & Cornwall, A. (2004), 'What is the "Rights-Based Approach" All About? Perspectives from International Development Agencies', IDS Working Paper 234, Brighton, Sussex: Institute of Development Studies.

Sen, A. (1999), *Development as Freedom*, Oxford: Oxford University Press.

ter Haar, G. (1992), *Spirit of Africa: The Healing Ministry of Archbishop Milingo of Zambia*, London: Hurst and Trenton, NJ: Africa World Press.

ter Haar, G. (2007), 'Religious War, Terrorism, and Peace: A Response to Mark Juergensmeyer', in Gerrie ter Haar and Yoshio Tsuruoka (eds), *Religion and Society: An Agenda for the 21st Century*, Leiden and Boston: Brill, pp. 19–27.

Tylor, E. B. (1958) [1871], *Primitive Culture*. Repr. as *The Origins of Culture* and *Religion in Primitive Culture*, 2 vols, New York: Harper.

Tyndale, W. R. (ed.) (2006), *Visions of Development: Faith-Based Initiatives*, Aldershot: Ashgate.

Various authors (2005), *Religion: A Source for Human Rights and Development Cooperation*, The Hague: BBO.

Various authors (2007), *Transforming Development: Exploring Approaches to Development from Religious Perspectives*, Utrecht: Knowledge Centre Religion and Development.

Zohar, D. & Marshall, I. (2004), *Spiritual Capital: Wealth We Can Live By*, London: Bloomsbury.

Internet resources

http://berkleycenter.georgetown.edu/
http://hdr.undp.org.hd/
http://www.icco.nl/
http://www.religion-and-development.nl/
http://www.spiritualcapitalresearchprogram.com
http://www.worldbank.org/

3
Muslim Shrines in Cape Town: Religion and Post-Apartheid Public Spheres

Abdulkader Tayob

This chapter is a close reading of public engagement over Muslim shrines (*kramats*) at Oudekraal outside Cape Town. These shrines are situated along a breathtaking drive on the Atlantic Seaboard of Cape Town, between the upmarket suburbs of Camps Bay and Llandudno. Between 1996 and 2007, the *kramats* were the subject of a legal battle and public debate between the state, the owner of the ground on which the *kramat* are founded, and various Muslim and environmental activists. In the post-apartheid public sphere the *kramats* were reconfigured as heritage sites located in environmentally sensitive areas. At the same time, it was also clear that the *kramats* revealed a Muslim public that extended beyond the boundaries and concerns of the national public sphere. This Muslim public was the product of intense contestation and diverse appropriation by various Muslim groups. This chapter argues that a Muslim public was invariably refracted through multiple engagements: with the state, among competing theological groups, and through religious activities and agencies that used symbols such as the *kramat* to create individual and social meaning.

The dispute over these *kramats* also provide an opportunity to closely examine the relationship between culture and the post-apartheid South African state. Since the process of democratization, the South African state has been balancing many competing demands. One of these is a constitutional respect for cultural rights, together with a wish to forge a new nation. The former responded to the history of apartheid and colonialism when certain cultures were considered inferior and denigrated in numerous ways. The celebration of all cultures formed an important part of the new South Africa. On the other hand, the cultivation and growth of these cultures represents a continued division of the country into ethnic, religious and cultural groups that inhibit the development of a national identity and rights-based citizenship. This tension between the two demands has inevitable consequences on a political level, as some recent research has shown. Cultural groups, particularly those marginalized in the past, enjoy the attention of South African politicians for votes and support (van Kessel & Oomen, 1997; Radhakrishnan, 2005).

The literature on culture and politics has quite correctly examined these issues in the light of the weaknesses and failures of the postcolonial state (Goldsmith, 2000). Within Africa, much of the literature has examined the extent to which ethnicity has been maintained and even strengthened in the distribution of goods and wealth (Sangmpam, 1993; Samatar, 1997; Weinreb, 2001). Bayart's politics of the belly has cast a long shadow on the meaning of culture in the African state. With respect to religion, however, there are some interesting differences worth examining more carefully. Haynes, Gifford and others have shown how religion has followed patterns established by the politics of ethnicity. They have pointed to the propensity of religious leaders to take maximum advantage of the political structure of the postcolonial state. Religious leaders have been willing to support those in office, in exchange for favours for their particular groups (Haynes, 1995; Gifford, 1996; Heilman & Kaiser, 2002). On the other hand, Ellis and ter Haar (1998) have been more radical in their analysis of the effect of religion on the state. They argue that the modern state has capitulated to religious vocabulary and meaning, pushing its modern dimension into the background. Examining the particular history of Muslims in African states after colonialism, O'Brien (2003) has offered a similar theory for the symbolic appropriation of Muslims of the African state. These scholars argue that religious symbols permeate and intrude into the structure of the African state.

In the light of these reflections, the *kramats* provide an interesting case study for examining the role of culture in general, and religion in particular, in the South African public. Closer attention to the detailed engagement at the *kramats* provides evidence that religion straddles the broad political space. On the one extreme are those who support the development of the public sphere and its national general goals. At the same time, however, the *kramats* produce networks of support and communities of meaning beyond the postcolonial state. These are not always antagonistic to the state, but neither do they merge with its interests. They demonstrate how people appropriate religious spaces to create meaning through various kinds of individual and social action.

The kramats at Oudekraal

The *kramats* of Oudekraal consisted of two sets of graves separated by approximately 500 metres. Each of these gravesites were dominated by one *kramat*, a shrine that has been marked as the grave of a saint by visitors and pilgrims. The first and most prominent of these was the *kramat* of Shaykh Noorul Mubeen. This was located within a rectangular room built in the 1980s by the Cape Mazaar Society. Established in 1982, the society embarked upon a programme of building and maintenance of all *kramats* in the Western Cape. With respect to this site, the society was approached by a group of men who wanted to build a shrine over the grave of Shaykh Nurul Mubeen. A load of bricks was donated and delivered to the site. For the next few years, visitors to the shrine were asked to take as many bricks as possible to the *kramat*. Ninety-nine steps were eventually built leading to the shrine. The site was divided into sections, each one with either well-trimmed lawn or vegetable patch. There were benches or tree stumps for visitors. It also had a cooking area for preparing meals on the death anniversary of the saint (usually in March or April). A mountain stream flowed into the site, and it was directed to a structure from which visitors could drink or perform ablution. The shrine did not have a dome, but provision seemed to have been made for one. The interior was painted white, decorated by an assortment of Qur'anic calligraphy prints, a plea for help to Shaykh Abd al-Qadir al-Jilani, photos of the mosques in Mecca and Medina printed by the Ministry of Hajj in Saudi Arabia, and a plaque giving the date on which the saint was brought to the Cape in 1713. There was also a large chest to receive donations, a number of bookshelves for prayer books, and a small area set aside for performing the Muslim prayer (*Ṣalāt*). The whole site was immaculately clean and well kept thanks to the dedication of a family in the Bo-Kaap that has maintained the site for the last three generations.

Sayyid Jaffer was the centre of the second set of graves. In 2009, I counted a total of 15 other graves in the vicinity, covered with brightly coloured cloths pinned by stones to the ground. The grave of Sayyid Jaffer was not set in a building, but covered by a steel canopy with green shades. There were seats attached to the structure, so that those performing the rituals of remembrance (*dhikr*) could sit around the *kramat*. Two large plastic tags indicated the name of the *kramat*, one had Sayyid Jaffer on it and the other Saint Jaffer. The *kramat* was adopted by a person called Yusuf Hawa from Cape Town, who had informed the Cape Mazaar Society that the saint did not wish to be fully covered. Mr Hawa had asked the Society to look after the *kramat* when he could no longer reach it due to old age. There were no stairs leading to this *kramat*, but some loose slabs of concrete on the lower level directed the visitor up the mountain. Higher up there were small pieces of cloth tied to the trees, strategically placed to encourage the visitor as he or she went up the mountain. Even though the *kramat* of Sayyid Jaffer was clearly identifiable by its steel structure, there was one particular grave further up the mountain that stood out in another way. This one had 40 sheets compared to the four over Sayyid Jaffer. It seemed that some people were venerating this *kramat* for some special reason.

These graves were located on land which had been owned by Oudekraal Estates (Pty) Ltd since 1965. The area was earmarked as commercial property in 1957, and township rights were confirmed in 1961 on graves around Sayyid Jaffer. In 1996, the owners presented engineering plans for developing an elite residential area on this particular portion, in line with other similar developments in the area. Neighbours and environmentalists were alerted to these plans, and soon raised objections with the City Council. The developers and owners approached the Cape Mazaar Society to explain their plans, hoping to get support from what they assumed to be a less active public group. They also offered to mechanically move the graves on portion seven to another suitable location. The Cape Mazaar Society was aware of their application to the City Council for developing the land. They found the offer from the developers attractive, but were wary of the sentiments of Muslims who visited the site. They called a meeting in a community centre in Rylands, a predominantly Indian area in Cape Town. The meeting attracted great interest, and included members of the People Against Gangsterism and Drugs (PAGAD) whose campaigns were just emerging in the city. The meeting expressed no desire to negotiate with the developers, and launched an ad hoc Environment Mazaar Action Committee to protest against the proposed development. A highly successful protest march was organized on 20 October 1996 on the slopes of Table Mountain, and sealed the fate of development as far as the Muslims and broader public were concerned.

In 2001, the owners of the site took legal action against the Cape Town City Council, the South African Heritage Resource Agency, and the National Parks Board. The City Council became the main body that responded to the appeal. The owner insisted that the City Council honour its decision to permit the establishment of a township on the property. In response, the City argued that the township rights given in 1957 had lapsed. In 2002 the matter was heard in the Cape Town High

Court and the judge awarded in favour of the defendants. The owner appealed against the decision and the case was heard in the Supreme Court of Appeal in 2004. Here, the court agreed with the Cape Town High Court but asked the City Council to formally set aside the earlier administrative decision of 1957 to approve the development of a township. In 2007 the Cape Town High Court formally acceded to this request. The legal process then sealed the fate of development.

Cultural rights in post-apartheid South Africa

Various courts dealt with the delays in the bureaucratic process, but more importantly reflected on the meaning of cultural rights guaranteed in the constitution. However, they eventually dealt with the substantial issue of the graves on the site. On this issue, the cultural rights of Muslims were affirmed in the context of a new constitution and the reality of marginalization of certain cultural practices and groups in the past.

The courts found that the various administrative bodies in the 1950s had simply ignored the graves. On the basis of evidence of both experts and Muslim testimonies they were satisfied that the site was regularly visited since 'time immemorial'. Aerial photographs taken in 1945, 1951 and 1986 had pointed to footpaths leading to these sites and land surveys during the hearings confirmed the existence of numerous graves. The evidence presented did not go earlier than the twentieth century, but the courts accepted the area as a site of slave and Muslim settlement – focused around literate, spiritual leaders who were presumably venerated after their demise – since the eighteenth century.

In constitutional terms, the courts confirmed the right of Muslims to practice their religion and culture. In his High Court Judgment in 2002, Judge Dennis Davis rejected the application for development, and favoured the 'cultural rights' that Muslims enjoyed over the area (Davis, 2002). The Supreme Court of Appeal also pointed out that permission had been granted for township rights with total disregard to the graves. Such a decision had ignored a common law principle against the desecration of graves and would go against the Bill of Rights in the post-apartheid constitution that guaranteed cultural and religious rights. In the final decision in the case in 2007, Judge van Reenen of the Cape High Court supported the cultural rights of Muslims to the site. He was critical of the inordinate delay of the City Council and the South African Heritage Association in completing the review process, but upheld the religious and environmental rights to these sites. He pointed out that the owners had done very little during this time to assuage the fears of Muslims that development would deny access to the site. The counsel for the developers, the judge said, focused more on the unreasonable delay of the government authorities and questioned the retrospective nature of the Constitutional support for cultural rights. He said that ignoring the environmental and religious rights 'would allow criminal conduct in the form of the desecration of graves'. Turning to a significant point in this case, he referred to the historical background of Muslims in the country:

a large proportion of the members of the Muslim community were previously socially, politically and economically disadvantaged because of the repressive and disempowering political policies of the past and, for that reason, have not been in a position to effectively assert and protect their interests.

(van Reenen, 2007, p. 46)

The desire to affirm the cultural rights of those marginalized in the past was an important element in the dispute. This particular point framed the public protest and also the meaning of culture in post-apartheid South Africa.

Muslim public at the kramats

The public discussion and debate, as well as the judicial decisions about cultural rights were clear, but tended to present the Muslim response to the *kramats* in monolithic terms. It is interesting that in the final judgment of 2007 Justice van Reenen referred briefly to the diversity among Muslims:

The endeavour on the part of the first respondent [Oudekraal Estates (Pty) Ltd] to question the correctness of the applicants' reliance on the abhorrence of Muslims to the exhumation of graves by having referred to incidents where it had been permitted in the past, floundered because it cannot be disputed that in terms of a *Hukem* issued in 1973, Muslim burial grounds are considered to be sacred and have been prohibited from being sold or the bodies buried there exhumed.

(van Reenen, 2007, p. 25)

The judge was referring to a declaration of the Muslim Judicial Council at the height of the apartheid Group Areas removals which stated that mosques and burial grounds were sacrosanct and could not be sold or used for purposes other than that for which they were originally intended.[1] The presence of a *hukem*, notwithstanding, there was a vibrant discourse surrounding the *kramats* that merits more careful analysis. While Muslims accepted the cultural rights supported by the constitution, they also engaged in a complex discourse that points us to the wider impact of religion and culture in society.

The Muslim discourse can be distinguished by at least three inter-related components. The first was clearly identifiable as Public Islam identified by Salvatore and Eickelman (2004) and Salvatore (2000), and others as an attempt to relate religious values to the common good of the public sphere. The second was an equally familiar debate about the *kramats* as innovations (*bidà*) and thus reprehensible to visit for any purpose. And the third component was a diverse appropriation of the *kramats* that described their popularity and which seemed to have impressed the court. Each of these components can be elaborated in their distinctive manifestation.

As the dispute became part of broader public debate in Cape Town, Muslims reflected on the meaning of the *kramats* beyond their religious value. Many accepted them as symbols of resistance in the history of colonialism and imperialism. One particular response is worthy of more careful observation, as it provides an excellent example of what many commentators on Islam and public life have referred to as Public Islam. Hassan Walele was a quantity surveyor and Qur'anic teacher who has emerged in Cape Town as a popular Sufi leader since the late 1990s. His statements, made at the demonstration of 1996and subsequently on radio, may be seen as a translation of a religious discourse into an ethical one that appealed to all the people of Cape Town. His main focus was not the spiritual value of the *kramats* but their broader public significance. Speaking of the 1996 protest, he mentioned the power of mass organization across religious divisions:

> Development was kept at bay because of the public outcry. When people spoke up, we had the voices of up to 80,000 people, across the cultural and religious lines, making this a South African, as opposed to a Muslim issue. That speaks of a strong public sentiment.

He went further, and proposed an ethic of Islam and environment:

> [...] It is one of the most endangered habitat types in the world with only 3% habitats like this remaining in the world. Some botanists

say it contains an unspoilt habitat for the Renosterveld plants and animals. It is the home to very rare, endangered butterflies and oil collecting bees which are almost extinct.

We cannot simply overlook these things and dismiss it as merely being nature when our own Quran gives us guidance on how to respect even the animal and plants.[2]

Imam Walele was, however, aware of the fact that not all Imams and religious leaders were supportive of this approach. It was, he said, 'time for the ulema to raise these issues on the environment and other current matters like poverty and crime from the mimbar [pulpit]'...'instead of restricting themselves to boring matters like halaal [the permissible] and haraam [the prohibited] from the pulpit' (ibid.). His was clearly a discourse that crossed the boundaries of religious identities to address the major ethical and socio-political issues of the day. This was a very good example of Public Islam that was produced by both specialists and non-specialists, presented to a common public, and forged notions of the common good (Salvatore, 2000; Salvatore & Eickelman, 2004).

Judging from Walele's remarks, the dominant ethos among religious leaders was to focus on purely 'religious' concerns, and only touch briefly on issues of the environment, poverty and crime. In this particular issue, public debate among Muslims was more clearly dominated by a juridical (*fiqh*) and theological (*kalam*) argument about the permissibility of visiting the shrines and venerating them. I would like to identify this as a second component of the Muslim discourse. The local Muslim radio station, the *Voice of the Cape*, also provided a platform for focusing on this issue. It invited a well-known religious leader Mawlana Taha Keeran to discuss the *kramats*. Keeran was a member of the *fatwa* committee of the Muslim Judicial Council and a principal of a Deobandi school of learning that would in principal be opposed to visiting shrines. Some of the callers lamented a decline in the interest in the *kramats* within the Muslim community, laying the blame for this on negligent parents who did not fulfil all their religious responsibilities and current theological disputes on the importance of *kramats*. One caller captured both sentiments in this statement:

We have to accept responsibility for neglecting our kramats. We all grew up visiting it at weekends and recognised its importance. I don't blame the youth if they don't know of its significance today when it is we who are adults who have failed them. However, it is when some of the ulema began to raise the practises of bidà at the kramats that people began to move away from this custom.[3]

Another caller concurred, but also pointed to the impact of reformist ideas from abroad:

> Many of our leaders, especially those who returned from Saudi Arabia spoke against visiting the kramats and that helped to sow if not dissent, then confusion. What is needed is more clarity on this issue.

The radio medium provided an opportunity for the dissemination of reformist ideas. However, with the telephones lines open to contestation, it also provided an opportunity for listeners to question and challenge theological positions. And the idea of history and heritage seemed to be inscribed in the questions. Mawlana Taha Keeran's response was varied and measured. He steered the callers away from the spiritual value of the *kramats*, but thought that the 'secular' public value of history and heritage were worthy of remembrance:

> If that kabr [grave] is part of our history, then it needs to be preserved and holds a certain sentimental value. However, some take it beyond sentimentality to a spiritual value, which is where people differ and one needs to hear both sides of the argument...

> If this was regarded as a heritage site, then it is Islamically permissible to visit it. Although we differ on the spiritual value of visiting it, there is agreement that such places should be respected. If this is regarded as part of a community's history, then visiting such places is even encouraged.

As an expert in Islamic law, moreover, he could not summarily reject the right of the owner. Here too, however, he conceded the public sentiment about the significance of the site:

> If he (the developer) is able to isolate the grave sites on the property and preserve it, I see no reason why he is not able to develop it. One cannot deny him that right. However, if there is proof that an entire community lived there and there are unidentified graves, then a case could be made for such a community to obtain rights to this site.

Clearly, the theological rejection of the *kramats* was put under the spotlight of a public discourse. History and heritage, both promoted by the post-apartheid state, were making their presence felt within the Muslim debate. While the Mawlana was clearly not in favour of promoting the *kramats* as religious sites and symbols, he could not reject the public demands made on him. Radio, particularly where listeners were invited to call in, provided an effective medium for the shaping of this new discourse, and the Mawlana seemed to embrace the historical and heritage value of the sites as an effective response to this public debate. Accepting the *kramats* as a symbol of heritage, exactly as promoted by the post-apartheid state, was the most effective way to respond to callers who were accusing him of undermining the historical record of Islam in the Cape.

It is clear so far that the site was of great importance to Muslims who responded with enthusiasm and in great numbers to the meetings and mass march, and kept the radio discussions alive with challenging questions and protestations, even against well-established religious leaders. In the following section of the chapter, I would like to further elaborate on this third component of the discourse. I will begin with legends told about the site, and then go on to the people who visit the *kramats*. All these inscribed a rich discourse that must be placed alongside the public discourses over values, theology and correct practice.

The *Guide to the Kramats of the Western Cape* was compiled by M. Jaffer, and related a number of local legends of how Shaykh Nurul Mobeen eventually came to this spot. All the legends connect him with a successful escape from Robben Island, the world-famous prison of apartheid opponents:

> According to a popular legend he escaped from Robben Island by unknown means and came to make his home in this desolate spot. Soon he made contact with slaves on the estate in this area teaching them, mainly at night, the religion of Islam. When he died he was buried on the site where he had most frequently read his prayers. After a time a wood and iron structure was erected around the grave, that acted as the first tomb.

> In a second legend, it is claimed that he swam from Robben Islam across the Atlantic Ocean and made good his escape. His tired body was discovered by slave fishermen. They nursed him to health and hid him on the mountain side, providing him with all his requirements. The fishermen soon discovered he was a holy man and started to take lessons from him. Sheikh Noorul Mubeen became their Imam and counselled them in their moments of difficulties.

His mountainside refuge, aside from allowing him to easily detect danger, gave him a magnificent vantage point from which he could see the towering peaks of the Twelve Apostles and the quiet dignity of Lions Head.

An alternative version is that he did not swim, but walked across the Atlantic Ocean from Robben Island to the mainland.

A present day legend tells of a spirit on horseback from Robben Island who still comes to take lessons from his teacher. He is seen, so they say, at about midday on a white horse coming across the ocean from Robben Island.

(Jaffer, 2004, pp. 30–1)

The legends were pregnant with meanings and allusions of the *kramat*'s location and power. They suggest a popular memory that did not respect the artificial boundaries between the religions, or the laws of the colony or of nature. The legends seem to weave Islamic and Christian narratives together. They also connect the site to the prison-island of the Dutch Company rule and apartheid, stitching together a narrative of Islam and resistance. In interviews conducted with Muslims who visited the shrine, they emphasized the fact that the saints had brought Islam to the Cape, and preserved it for posterity against great odds and obstacles.

In interviews conducted with members of the Cape Mazaar Society, the society that has facilitated the building of the shrines and has given continued support to visitors, I discovered other aspects of the *kramats*. Most of the visitors seemed to be of Indian origin, but it is interesting to note that they have not been restricted only to Muslims. I was, however, told that the Senegal Tijani Shaykh, Hassan Cisse, confirmed the spiritual eminence of this *kramat*. Within the tradition of visiting the *kramats*, the *kramat* of Nurul Mobeen is known for providing an opportunity to obtain spiritual insight or solace. Tuan Jaffer, on the other hand, attracts those seeking healing. Visitors regularly spend long vigils from midnight to the early hours of the morning to attain physical health, or protection from ailments.[4]

The Cape Mazaar Society guided me to another group who organized the *Urs* celebration of the saint, Nurul Mobeen. The *Urs* was a death anniversary of the saint, popular in South Asia, but the word referred to a marriage, and signified the union or meeting of the saint with God. In 2009 this celebration took place at the Nurul Mobeen shrine on 29 April. The organizers I met had inherited the tradition from their parents. The celebration consisted of a *dhikr* of the Qadiri Sufi order, and explained the prominence of a prayer addressed to its great founder in the shrine. The *dhikr* was attended by about 250 men and women. The men filled the shrine building slowly throughout the morning, while the women sat outside. A few women came close to the door, but did not venture beyond the threshold. The *dhikr* was recited in Arabic from printed booklets. The highpoint was a rhythmic beating of tambourines, while a small group of men moved in a counter-clockwise direction around the grave. At this point the *dhikr* was recited in Urdu. Throughout the morning as devotees trickled in and took their places some brought sheets of cloth and placed them above the saint's grave. Towards the end the sheets were opened, passed over the heads of the visitors, and then spread over the grave. Bottles of perfume were also emptied in containers that were then passed around among the visitors.

At the meal after the event, I spoke to a number of people there, some of whom I had known as students at the University of Cape Town. They revealed that the *Urs* have been taking placing regularly for longer than anybody present could remember. However, nobody knew anything in particular about the saint himself. The date for the death anniversary was set by someone at the Cape Mazaar Society, but the devotees themselves performed an *Urs* that was connected to their own Sufi circle in the Indian suburb of Rylands. They held regular *dhikr* meetings at their homes and came to Shaykh Nurul Mobeen for the *Urs*. The spiritual tradition was not the only focus of the gathering. On my arrival at the *Urs*, I was told that the celebration was attended by all Muslims. One of the organizers emphasized that it was attended the previous year by a busload of Muslims from the townships, meaning Muslims of indigenous African origin. My conversations after the *dhikr* revealed the reason for this emphasis. The meeting was open to all, but it particularly attracted a sub-clan of Muslims of Indian origin. Kokanee-speaking Indians constitute the majority of Indian Muslims in Cape Town, and are further subdivided into a number of clans. Among them, the Habsanids gathered at the *Urs* of Nurul Mobeen. For young and old, the *kramat* of Nurul Mobeen provided a social axis for the identity of the group.

The *kramat* was also attended by smaller groups and individuals. I met two men and a woman, who told me that their teacher had recommended that they visit the *kramats* throughout the city. They made a habit of visiting a different one on a weekly basis. They also enjoyed the beauty and tranquillity of this particular site. The woman also emphasized 'power' that was evident for her at the site. She recalled the march against the development in 1996 in which she had participated and recalled how she had stood together with 'everybody', arm in arm, Muslim and Christian, men and women. I followed their trail to the religious leader and mosque that appeared to give them a strong sense of identity with the *kramats*, and that connected them to the shrines. I met Shaykh Salamatdat at the Beacon Valley mosque in a working-class area in Mitchel's Plein. He manages a thriving centre. Shaykh Salamatdat was then thirty-five years old, and had studied at the Deoband seminary in Newcastle, Kwazulu-Natal, and later in Lucknow, India. His father was closely connected to the Chisti Sufi order, from which he personally had drifted away when he joined the reformist Deobandis. On his return from Lucknow, he was introduced to a Sufi order from Jordan, and appointed its representative (*muqaddam*) in South Africa. Since then the mosque has become the centre of a vibrant Sufi movement.

There are probably more groups connected to the shrine. But there were also a number of individuals who were not directly connected to any organized group. I met a man who came on his own to perform a special devotional prayer (*salat al-tasbīh*) that demanded considerable concentration and time. He had been introduced to the shrine by his wife whose Malaysian father had settled in the Cape in the 1940s. The latter had made a habit of visiting the shrine and bequeathed the practice to his family. Then there was a couple with their two children, a visually impaired four-year-old daughter and newborn baby. The man went inside the *kramat* with his daughter and then his newborn, but the woman stayed outside. In a short discussion afterwards, they told me that they were from Durban (Kwazulu-Natal) and had intended to visit the *kramat* for some time. The young man was an accountant. The woman told me that she did not want to go in out of respect for the saint. They were careful to tell me that they did not address the saint for assistance. I met another couple who came to the *kramat* of Tuan Jaffer on a shiny motorbike. They too had decided to visit the *kramat* on the advice of a local Imam who emphasized the historical memory of the Muslims. They seemed not to know too much about the *kramat*, and to have given up on finding Tuan Jaffer when I met them. However, they told me of earlier experiences of visiting the shrines with their families. I met other individuals who liked to recite a portion of the Qur'an at the *kramat*. One of them recalled that he used to go at night after the evening prayer. Another told me that he was introduced to the *kramat* by his father-in-law who had placed a cloth over Tuan Jaffer soon after their engagement. He himself no longer visited the *kramats*, and told me that the tradition died out with the passing of his father-in-law.

Reflections on Public Islam and the post-apartheid state

The material presented here provides a good opportunity to closely examine the nature of Public Islam, and the role of religion in South Africa's democratic politics and public sphere. Bringing them together provides a basis for examining Public Islam within a democratic ethos that supported the common good of cultural rights. My contention is that Public Islam was clearly influenced by its relation with the state and its publically determined goals for the common good. However, Public Islam was significant for religious life beyond the state. It provided a basis of organization, mobilization and social meaning for diverse cultural goals and activities. The state responded positively to the social capital demonstrated by Muslims and environmentalists. Their ability to mobilize opinion and mass gatherings may have forced the state to take note.[5]

A perceived threat to the *kramats* drew Muslims, environmentalists and the democratic state institutions together. Spurred by different interests, the idea of cultural rights provided a very clear ground for the articulation of a common good. The court confirmed these rights, and mass protests articulated them in public practices. Moreover, public attitudes and feelings also led to a change in and a redefinition of the *kramats*. Standing in the centre of a theological dispute, the *kramats* were reconfigured with different values. Under the impact of the new political dispensation, a new public ethic was promoted. The role of the media (radio) and non-specialists were instrumental in the formation of Public Islam. To this extent, Public Islam in Cape Town conformed to the definition of Eickeleman and Salvatore:

> [...] highly diverse invocations of Islam as ideas and practices that religious scholars, self-ascribed religious authorities, secular intellectuals, Sufi orders, mothers, students, workers, engineers, and many others make to civic debate and public life. In this 'public' capacity, 'Islam' makes a difference in configuring the politics and social life of large parts of the globe, and not just for self-ascribed religious authorities.
>
> (Salvatore and Eickelman, 2004b, p. xii)

Public Islam produced a space where the boundaries between specialists and non-specialists collapsed. Such a shared platform significantly altered the goals of religious discourse, according to Salvatore, from the 'articulation of a cosmological order' to 'the common good to be argued about' (Salvatore, 2000, p. 15). Walele was a good example of some-one who contributed significantly to the articulation of Public Islam in this case. However, Mawlana Taha Keeran's capitulation was even more interesting, as it illustrated the transformation under public pressure, armed now with a medium (radio), a political context (democracy) and a discourse of cultural rights.

My contention in this debate, however, is that there is wider sig-nificance to the *kramats* than that conveyed by the concept of Public Islam. Salvatore in particular was focused on the decline of theological debates over public issues, while many others seemed equally concerned about Public Islam's potential to produce progressive and democratic values in Muslim societies. The dissenting voices in this debate have emphasized the religious values promoted within Public Islam, or Public Islam's global discourse which was insulated from the secular, demo-cratic terms of the debate. Hirschkind, representing the former, has written of Islamic counter-publics in Egypt where conformity to reli-gious values and ideals gave little room for the deliberation of common values and the common good (Hirschkind, 2001, 2006). In a different vein, Bowen has pointed to a global Islamic discourse that dominates Public Islamic debates, allowing little room for the intrusion of local values and terms to be incorporated therein (Bowen, 2003). Both these arguments have raised important critical questions about a hasty judge-ment on the convergence of Public Islam and the public sphere in any given context. However, their analysis does not include the diversity of voices and contributions that make up public engagement of religion.

My closer examination revealed that this broader engagement with the *kramat* went beyond the articulation of Public Islam and its confor-mity with the post-apartheid public sphere. The *kramats* were managed, built and visited by a variety of other individuals and social groups. In search of healing, spiritual solace and community, they developed a wide-ranging set of practices, strategies and networks around the *kramats*. Many of these practices were directly manifest at the graves, but their impact and effects extended beyond these sites. When exam-ining this broad field of activities held together by the *kramats*, we notice a range of discourses and subjectivities. These include the emer-gence of an organization such as the Cape Mazaar Society that has managed to define and inscribe the *kramats* with practices, memories and also a future. They include the networks of ethnicity and spiritual adepts that orbit around the *kramat*. Equally importantly, they include multiple meanings of belonging, place and time for individuals and groups alike.

Such social capital must be put in the context of thinking about religion and about the post-colonial state. As I mentioned in the beginning of this chapter, much has been written on the relationship between the state and cultural brokers. The postcolonial state imposes its logic on cultures, or vice versa. The public debate over the *kramats* pointed to a different scenario. Some elements of Public Islam conformed to the rights-based constitution, adjusting its values to those of the constitution. But this was not simply an imposition of the state, but the way in which ordinary, lay Muslims used the language of the constitution to force some adjustment in the articulation of a Public Islam. By itself, this was an interesting development, and worthy of reflection in the broader discussion of Public Islam. However, the *kramats'* sites were not exhausted by the political culture of rights. Their cultural production extended beyond the state and beyond its ethics. Embracing social and psychological goods, the *kramats* represented a cultural resource that was attractive to many.

Notes

1. In Islamic law, there was a difference of opinion whether mosques and cemeteries may be dismantled or used for other purposes, once they had collapsed through disuse, neglect or the passage of time (Jād al-Haq, 1983). In an attempt to prevent this disagreement from being exploited by the apartheid regime, the MJC issued this statement (2007).
2. 2007. No Way on Oudekraal: EMAC. http://www.vocfm.co.za/public/articles.asp? Articleid=25703 (accessed 4 April 2007).
3. 2007. Oudekraal: Strong Support, But Clarity Needed. http://www.vocfm.co.za/public/articles.asp?Articleid=25745 (accessed 4 April 2007).
4. Interview with Mahmud Limbada and Khaleel Ali at the Cape Mazaar Society offices, 3 March 2009.
5. There was one other contentious gravesite closer to the city at Prestwich, uncovered during excavations at a building site in 2003. Protests and demonstration did not cease bring the city to the support of the underclasses in the same way that it responded to the Oudekraal. City leaders may have been persuaded by the social capital demonstrated by Muslims at Oudekraal (Jonker, 2005; Sheperd, 2007).

References

Bowen, J. R. (2003), 'Beyond Migration: Islam as a Transnational Public Space' *Journal of Ethnic and Migration Studies*, 30 (5), pp. 879–94.
Davis, J. D. (2002), 'Oudekraal Estates (Pty) Ltd V. the City of Cape Town and Others' *High Court of South Africa*, Cape Town: Cape of Good Hope Provincial Division.

Ellis, S. & ter Haar, G. (1998), 'Religion and Politics in Sub-Saharan Africa' *The Journal of Modern African Studies*, 36 (2), pp. 175–201.

Gifford, P. (1996), 'Christian Fundamentalism, State and Politics in Black Africa', in D. Westerlund (ed.), *Questioning the Secular State: The Worldwide Resurgence of Religion in Politics*, London: Hurst & Company, pp. 198–215.

Goldsmith, A. A. (2000), 'Sizing Up the African State' *The Journal of Modern African Studies*, 38 (1), pp. 1–20.

Haynes, J. (1995), 'Popular Religion and Politics in Sub-Saharan Africa' *Third World Quarterly*, 16 (1), pp. 89–108.

Heilman, B. E. & Kaiser, P. J. (2002), 'Religion, Identity and Politics in Tanzania' *Third World Quarterly*, 23 (40), pp. 691–709.

Hirschkind, C. (2001), 'Civic Virtue and Religious Reason: An Islamic Counter-public' *Cultural Anthropology*, 16 (1), pp. 3–34.

Hirschkind, C. (2006), *The Ethical Soundscape: Cassette Sermons and Islamic Counterpublics*, New York: Columbia University Press.

Jād al-Haq, Alī Jād al-Haq (1983), 'Fatwā'. In Al-Fatāwā Al-Islāmiyyah Min Dār Al-Iftā Al-Misriyyah, 26: 3259–61. Cairo: Jamhuriyyat Misr al-ʿarabiyyah: Wizārat al-awqāf, al-majlis al-aʿlā li-shuʾūn al-islāmiyyah.

Jaffer, M. (2004), *Guide to Kramats of the Western Cape*, Cape Town: Cape Mazaar Society.

Jonker, J. (2005), 'The Silence of the Dead: Ethical and Juridical Significances of the Exhumations at Prestwich Place, Cape Town, 2000–2005', M.Phil dissertation, Cape Town: University of Cape Town.

O'Brien, D. B. C. (2003), *Symbolic Confrontations: Muslims Imagining the State in Africa*, London: Hurst & Company.

Radhakrishnan, S. (2005), ' "Time to Show Our True Colors": The Gendered Politics of "Indianness" in Post-Apartheid South Africa' *Gender & Society*, 19 (2), p. 262.

Salvatore, A. (2000), 'Social Differentiation, Moral Authority and Public Islam in Egypt: The Path of Mustafa Mahmud' *Anthropology Today*, 16 (2), pp. 12–15.

Salvatore, A. & Eickelman, D. (eds) (2004), *Public Islam and the Common Good* vol. 95, *Social, Economic, and Political Studies of the Middle East and Asia*, Leiden: Brill.

Salvatores, A. & Eickelman, D. (2004b), 'Preface Public Islam and the Common Good' in Public Islam and the Common Good (eds), A. Salvatore & D. Eickleman, Leiden: Brill, pp. xi–xxv.

Samatar, A. I. (1997), 'Leadership and Ethnicity in the Making of African State Models: Botswana Versus Somalia' *Third World Quarterly*, 18 (4), pp. 687–708.

Sangmpam, S. N. (1993), 'Neither Soft Nor Dead: The African State is Alive and Well' *African Studies Review*, 36 (2), pp. 73–94.

Sheperd, N. (2007), 'Archaeology Dreaming: Post-Apartheid Imaginaries and the Bones of the Prestwich Street Dead' *Journal of Social Archaeology*, 7 (3), pp. 3–28.

van Kessel, I. & Oomen, B. (1997), ' "One Chief, One Vote": The Revival of Traditional Authorities in Post-Apartheid South Africa' *African Affairs*, 96 (385), pp. 561–85.

van Reenen, J. (2007), 'The City of Cape Town and Others Vs. Oudekraal Estates (Pty) Ltd and Others', *High Court of South Africa*, Cape Town: Cape of Good Hope Provincial Division.

Weinreb, A. A. (2001), 'First Politics, Then Culture: Accounting for Ethnic Differences in Demographic Behavior in Kenya' *Population and Development Review*, 27 (3), pp. 437–67.

Radio Broadcasts

(2007), 'Hard Lesson Led to Oudekraal Fight Back: Mjc', http://www.vocfm.co.za/public/articles.asp?Articleid=25812 (accessed 5 March 2008).

(2007), *No Way on Oudekraal: EMAC*, http://www.vocfm.co.za/public/articles.asp?Articleid=25703 (accessed 4 April 2007).

(2007), *Oudekraal: Strong Support, But Clarity Needed*, http://www.vocfm.co.za/public/articles.asp?Articleid=25745 (accessed 4 April 2007).

4
Remaking Society from Within: Extraversion and the Social Forms of Female Muslim Activism in Urban Mali

Dorothea E. Schulz

Introduction

In January 1999, during a visit to Nara, a small Sahelian town situated in proximity to Mali's border with Mauretania, I made a surprising discovery. I had been familiar with this town between 1987 and 1990, when I had spent seven months in a small village located nearby. As I drove along the main road that connects this dusty town to Mali's capital, Bamako, I was stunned to see hundreds of women, some of them in all-white or all-black attire, others in colourful robes, but all of them covered with an additional prayer shawl, walking along the streets and scattered on the area close to the market square, the site of the weekly Monday market. Knowing that this was the third week of the holy month of Ramadan, I should not have been surprised by this overwhelming impression of female Muslim religiosity. After all, this town's location and population, and its geographical closeness to Gumbu, an older centre of Muslim trade and erudition, placed this area firmly on the map of Malian Muslim history. Still, what struck me as something distinct from the celebrations of the Ramadan I had witnessed in 1988 and 1990, were the number of piously dressed women; some of them sitting on mats around the central square, chatting and conducting commerce, while others, at a greater distance, were dancing and singing praise songs on behalf of the prophet Muhammad and his family. Also new were the various billboards announcing the presence of foreign Islamic associations, and the enormous array of electronic devices, such as loudspeakers, microphones and tape recorders, that generated an overwhelming cacophony of images and sound bites of religiosity. Still trying to digest all these impressions, I overheard one of my travel companions, clearly annoyed, say to his neighbour: 'Now you can see what has changed in Mali! Nothing against celebrating Ramadan in public, this has always been our heritage. But these "religious women", they knock me off. They imitate an Islam that is foreign to us. Why do they need to make such a show of their faith? Ultimately, all they do is politics.'

Mali's urban landscapes have undergone a remarkable transformation over the last decades. Since the demise of single-party rule in 1991, and the subsequent transition to multiparty democracy, emblems and idioms of Muslim piety have gained unprecedented public visibility, along with a mushrooming infrastructure of Islamic proselytizing (Arabic, *da'wa*) in the form of Muslim associations, mosques and schools (*medersas*[1]). These developments had started already in the early 1980s, facilitated by significant donations from charitable organizations and individual and public sponsors from the Arab-speaking world; yet they took on a new dynamic and intensity after the political opening of the early 1990s. They are associated by many Malians with 'Arab Islam', by which they refer to practices and interpretations of Islam that draw inspiration from early twentieth-century Salafi-Sunni reformist trends in Egypt and Saudi Arabia. A distinctive feature of the Islamic renewal movement is the prominence of women who present themselves as articulators and icons of moral reform. They maintain that the ills of Malian society can be remedied only by returning to the original teachings of Islam, and by inviting others to join their search for a 'greater closeness to God' (Schulz, 2008b). These women refer to themselves simply as 'Muslim women' (*silame musow*) and thereby distance themselves from 'other women' (*musow to*), who, they imply, are not 'real Muslims'. A principal marker of difference, 'Muslim women' claim, is that 'other women' have not decided to 'don the veil' (*ka musoro siri/ta*), a dress practice their critics denounce as a sign of 'Arab Islam's' unfortunate influence on local Muslim practice.

The 'Muslim women' organize themselves in neighbourhood groups (singular, *silame musow ton*, literally 'Muslim women's group') to 'learn to read and write' the Qur'an and to engage in joint religious practice. Their leaders, referred to as 'group leaders' (singular, *tontigi*) or with the honorific title *hadja*, offer advice on proper female conduct in ritual and daily settings. Some *hadjas* broadcast their teachings on audio-recordings and on some commercial local radio stations that have emerged in urban areas since the early 1990s.

As intimated by my travel companion whom I quoted above, the religious activities of 'Muslim women' in public arenas are not new. Relatively novel, however, is that the Muslim women invest them with a new significance. They stress women's key role in societal renewal and publicly articulate their conviction that women's God-ordained role relegates them to the domestic realm. This parallel emphasis on a woman's domestic and public responsibilities reveals their distinctive conception of the significance of personal ethics to collective well-being and public affairs. Still, although the Muslim women and their groups intervene in the same social realms as Western donor-funded non-governmental organizations (NGOs), representatives of these NGOs and of the state deny these Muslim activists equal membership in civil society. This denial comes out in many instances, such as during public debates of a family draft law between 1999 and 2001 (Schulz, 2003). Although some Muslim associations had been invited to participate in these debates, their positions and arguments were persistently sidestepped by state officials and NGO leaders who considered the Muslim activists representatives of 'religion', of a 'radical' foreign (that is, 'Arab') influence, and thus a threat to the secular state. They did not concede them a role as representatives of a particular vision of the common good in an emergent, heterogeneous 'civil society'.

The views of these 'secularist-minded' critics are echoed by many studies on the political role of religion in African state politics of the 1990s, studies that establish an – often implicit – contrast between Christian organizations and those created by Muslims. Bayart and Gifford, for instance, highlight the anti-authoritarian potential and progressive contribution of (mostly mainline) Christian churches in bringing about democratic change (Bayart, 1993; Gifford, 1995, 1998). Recent explorations of 'faith-based development' in Africa, too, focus attention on Christian initiatives (e.g. Bornstein, 2003; Hofer, 2003). Muslim organizational forms, in contrast, although they resemble Christian groups in their institutions and conventions of socializing, are viewed as promoters of 'religious' values and disregarded in their socially constitutive role.[2] Traces of this interpretational scheme are evident in recent studies on Muslim movements in Africa that centre their attention on the question as to whether Muslim movements challenge the secular nation state (e.g. Otayek, 1993; Brönig & Weiss, 2003; Soares, 2006; Soares & Otayek, 2007). While these studies ultimately propose a different interpretation, they illustrate that efforts to 'redeem' Muslim movements from the charge of threatening democratic politics risk reproducing the analytical frame they set out to refute.

What needs to be emphasized is that in much of Muslim Africa, religious networks and clientilism have historically operated not as a threat to secular state politics, but as part and parcel of its emergence (e.g. Villalon, 1995; Kane & Triaud, 1998). Also, Muslim activists' unanimous call for a 'return' to authentic Islamic values might conceal the highly heterogeneous character of their positions (Schulz, 2006b, 2007). The question, then, is what specific relationship of coexistence, contestation, and co-optation Muslim activists and organizations have established historically with the state. Only by understanding how Muslim interest groups insert themselves into state politics, will we be in a position to identify the reasons why 'religion' has gained in appeal and prominence in African nation state politics since the political liberalization of the late 1980s (e.g. Gifford, 1995, 1998; Ranger, 2003; Meyer, 2004; Fourchard et al., 2005). Questions we need to address concern the internal dynamics of these organizations; how they relate to transnational connections and influences; and how their ties to state institutions, and the logic of affiliation between group leaders and ordinary members, affects the terms on which they make their influence felt in public (see Joseph, 1997). We should also take account of continuities between these groups and their historical predecessors, and of the local dynamics that facilitated the rise of these associations at this particular historical juncture.

This chapter contributes to debates on the public role of religion in contemporary African state politics by exploring the significance that female patterns of Muslim sociality and renewal have acquired in Mali. Female religious associations and learning groups have been mushrooming since the mid-1980s. To date, they are among the most pervasive and visible structures of the moral reform movement. Given the scope and significance of these structures to Muslim women's self-understanding and to society at large, in Mali and elsewhere in the Muslim world, it is surprising that they have not received sustained scholarly attention. Many studies seem to content themselves with the view of female Muslim activists as pawns in the hands of male leaders, and largely leave aside the concerns that may guide these women.[3] A few recent studies depart from this earlier trend, yet focus primarily on the perspective of young, that is, unmarried, women (Le Blanc, 1998, 1999; Augis, 2002; but see de Jorio, 2009). With the exception of Halidou (2005), they say little about the organizational forms in which these Muslims engage, and that allow them to contribute to the re-articulation of state–society relations. Finally, although recent studies stress the importance of reformist influences from other parts of the Muslim world, their empirical focus is still on the national or local level.

This chapter explores the social institutional basis and political locations of the reform movement, and pays particular attention to the forms of sociality in which female supporters of the movement ground their aspirations. It argues that Muslim women's networks form part of a broader set of a religious institutional infrastructure and, similar to organizational forms run by Muslim men, intervene importantly in the remaking of state-society relations. The workings of these structures of female religious sociality are to be understood, at least in part, against the backdrop of transnational fields of influence, inspiration and material support. They can therefore be seen as the structural equivalent to the international and national Islamic non-governmental organizations whose activities, orchestrated mostly by Muslim men, are now starting to be explored by a few scholars (Weiss, 2002; Kaag, 2007). By retracing how Muslim women's organizational structures emerge in a triangle between transnational Muslim influences, institutions and discourses of Western aid agencies, and different local interpretations of Islam, the chapter elucidates both the stakes and the limitations of the women activists' 'politics from below'.

Bayart's notion of 'extraversion' opens up interesting lines of inquiry into the ways the moral reform movement's internal dynamics are generated in a translocal field of influences (Bayart, 2000). Several scholars have recently taken up this notion to make sense of the spectacular thriving of some Christian churches, old and new, in the current era of globalizing religious idioms, media, images and institutions, and to explain people's motivations to join these groups (Marshall Fratani, 1998; Englund, 2003; Meyer, 2004). African Christians' strategies of extraversion, these authors argue, are importantly geared towards accessing financial and institutional support generated along transnational connections, and thus form part of a local repertoire of responding to degrading economic conditions.

'Extraversion', read along these lines, sheds light on certain dimensions in the operation of 'transnational religion', and helps move beyond accounts, so prominent in the popular media since 11 September, that conflate different forms of transnational Muslim financial support by portraying them as carriers of 'radical' Arab influence. Yet, the notion of 'extraversion' needs to be further nuanced to make sense of new processes of internal differentiation, for instance along gender and age divisions, that occur within religious movements at local and national levels, and that often result from their insertion

into transnational structures of economic enterprise and spiritual salvation. Such a perspective on practices of extraversion illuminates
how Malian Muslim women's and men's differential insertion into
networks of religious patronage establish and reproduce relations of
inequality not only between men and women but, most likely, *among*
women.

Re/making society and self: Muslim women's moralizing endeavour in history

For an analysis of the significance of Muslim women's public
interventions, and of the particular forms through which they seek public visibility, the notion of 'public' offers a good entry point. Given
the historically specific legacy of state-society dynamics in postcolonial Africa, the relevance of this concept, developed with regard to
eighteenth-century Western Europe needs to be scrutinized.[4] As critics
maintain, Habermas's discourse-centred conceptualization of 'public-
ness' (*Öffentlichkeit*) in eighteenth-century Western Europe is based on
the universal (male, middle-class) subject whose critical argument is
predicated on the relegation of particularistic interests to the private
realm.[5] This chapter departs from the normative underpinnings of
Habermas's account, and instead uses 'public' as a heuristic tool to analyse Islam's new public manifestations in Mali and, more particularly,
the gender-specific, public encodings of Muslim virtue. Emphasizing the
historicity of women's religious engagements in Muslim West Africa, the
chapter interprets Muslim women's organizations in contemporary Mali
as forms of activism that, while drawing on long-standing historical
roots, take on novel forms and sociopolitical salience in the current era
of nation-state politics. It views the leaders and 'ordinary' members of
the Muslim women's groups as religious brokers whose contribution to
the spread of a new conception of Islam in contemporary Mali consists
primarily in their social and organizational skills and less in an input
into doctrinal debate.

Present efforts by Muslim activists, women and men, to articulate
norms of conduct in accordance with Islamic principles, and to make
them binding for all Malians as a matter of 'collective interest', draw
on earlier patterns of interaction between actors and institutions of the
state on one side, and different groups of Muslims on the other. In
contrast to the emergence of an influential religious elite backed by
the colonial state in Senegal and Northern Nigeria, the co-optation of
certain Muslim religious authorities and intellectuals by the colonial

administration in Mali rarely allowed these leaders to occupy positions of brokerage and influence in the colonial and postcolonial period
(Brenner, 2001). The repercussions of these historical terms of interaction between Muslim leaders and state officials and politicians, are
palpable in contemporary politics; they shape the aspirations of supporters of Islamic moral renewal and the particular forms of public
intervention they choose. The colonial legacy is also palpable in the
ways in which various Muslim leaders and interest groups express their
disagreements over doctrine and ritual praxis through a discourse on
'truth and ignorance' that has a long tradition in this area of Muslim
West Africa (Brenner, 1993a, 1993b; Soares, 2005; Schulz, 2006a). The
concerns addressed by supporters of Islamic moral renewal, especially
their emphasis on the civilizing effects of Islamic learning, are in line
with earlier reformist trends that date back to the late 1930s, when
Muslim intellectuals returned from extended stays in Egypt and Saudi
Arabia and initiated various reforms to counter the effects of French
colonial administration and education.[6] They sought to facilitate believers' access to the written sources of Islam, to deepen ordinary believers'
religious knowledge, and to purify conventional religious practices from
what they considered unlawful innovations (*bida'*). Whether they drew
inspiration from Salafi-Sunni influences in Egypt or from Wahhabi
thought in Saudi Arabia (Loimeier, 2003), they challenged the sources
and representatives of established religious authority (Brenner, 2001).[7]
In so doing, the reformists contributed to Muslim reasoning and activity
in ways that, regardless of local variations in forms and kinds of leadership, characterized social and political transformations throughout
Muslim West Africa (Kaba, 2000).

Starting in the late 1970s, a new generation of Muslim reformists,
eager to expand the local infrastructure of Islamic education and welfare, benefited from funds provided by a transnational *da'wa* movement
under the tutelage of Saudi Arabia (Brenner, 1993b; Otayek, 1993).
President Moussa Traore, then ruling the country, sought to control
the influx of money, and the attendant blossoming of an infrastructure of Muslim religiosity, through a two-pronged strategy. He organized
Muslim interest groups into a national association (the AMUPI[8]) closely
monitored by the state, yet simultaneously granted these groups special privileges, such as extended broadcasting periods on national radio
and television.[9] From the 1980s the funding provided by international
sponsors in the name of *da'wa* began to diminish considerably. But the
political changes during approximately the same period, initiated by the
ouster of President Traoré in 1991, opened up new spaces for Muslim

activism (Schulz, 2003, 2006). The contemporary Malian field of Muslim reasoning and debate continues to be characterized by a variety of interpretations of Islam and relations to the state administration and the government of President Amadou Toumani Touré (in office since 2002). Central to these dynamics are Muslim leaders of different backgrounds and pedigrees who mobilize followers through networks of patronage and financial support, and thereby contribute to a complex structure of religious and political patronage, which constitute various points of articulation between society and state (Schulz, 2004).

Muslim women's organizations form an integral part of this dynamic. Before situating their ethical endeavour within this field of Muslim debate and activism, let us take a closer look at the central concerns these women formulate.

Piety, publicity, representation: the objectives of female moral renewal

On a hot and dusty afternoon in May 2006, I paid a spontaneous visit to Kadiatou, a founding member of a Muslim women's association in Maniambougou, a popular neighbourhood in Bamako, with whom I had been in close dialogue since 2002.[10] As I entered her courtyard, I found Kadiatou and two visitors, both fellow members of the Muslim women's neighbourhood group, in deep conversation. The central topic of their conversation, as I gradually learned, was a woman (whom I will name Fanta) who had recently come to attend their group meetings and who, in spite of what Kadiatou and her friends considered a somewhat inappropriate demeanour, had expressed strong interest in becoming a group member. Kadiatou and her friends felt visibly torn about this request, and could not quite agree on what to do with this woman. Both visitors detailed scenes in which Fanta's 'pompous dress',[11] her habit of wearing jewellery, and her self-assertive, loud and 'mindless' talk, such as her constant gossiping about her neighbours, had irritated them. It was clear that Kadiatou shared this perception, and similarly wondered about Fanta's interest to join their group. 'But', she pointed out to the (approving) murmur of her two friends

> this is what our group is about, this is what our attempts to practice humanity (*hademadenya*) are for. We want to change the current state of social ills; we want to remake society in ways that obligation and respect rules once again our social relations. This is what God demands from us. [. . .] Many people mock us. Rather than criticising

women who do not behave properly, we have to educate ourselves
to be patient with these women and to show them the path to God
through our own example. For this to happen, we have to remind
ourselves of the true teachings of Islam. We have to teach ourselves
what the holy scriptures tell us, so that we can convince others
who call themselves Muslims but who are in the state of ignorance
(*jahiliyya*) because in their daily lives, they resist God's calling and his
eternal will.

Kadiatou's remark forcefully brings home the conviction, shared by
many Muslim women, that their moral quest requires a return to an earlier Islamic moral order. This perception stands in tension with the fact
that until the 1920s, Muslim identity and religiosity, although enjoying
a certain presence in many urban areas of Mali's south, has turned into
the religion of the majority of the urban (and rural) population only during the course of the colonial period. The renewal movement thrives in
urban areas where, historically, lineages associated with Muslim erudition did not have a stronghold. Hence, the identity as 'proper' Sunni
Muslims these women claim for themselves does not entail a return
to an older, original form; nor can many of their practices be seen as
a perpetuation of traditional Muslim religiosity. Kadiatou's mention of
'educating oneself' illustrates Muslim women's emphasis on their individual responsibility for personal salvation and for societal renewal.
Kadiatou also mentions that a woman's daily effort to become a proper
Muslim may bring her into conflict with her in-laws and her own
family. The emphasis on individual responsibility is noteworthy. It signals important changes in the significance of religious observance since
the colonial period, from an element of family affiliation, professional
specialization or of 'ethnic' identity, to an understanding of Muslim
religiosity as a matter of personal conviction. Kadiatou maintains that
becoming a proper Muslim requires that one 'understands' the Islamic
scriptures. This emphasis on direct access to Islam's foundational texts is
also illustrated by the fact that Muslim women present as primary rationale of their socializing activities the objective to 'learn', and refer to
their associations as 'learning groups' (singular, *kalani ton*). Their preoccupation with textual understanding reflects on changes in the field of
Muslim knowledge and interpretive authority. Until the 1970s, religious
learning and sustained ritual knowledge were limited mostly to women
of elite background (Sanankoua, 1991). As most of the 'Muslim women'
are young and from the middle and lower-middle classes, their learning
efforts demonstrate the widening of access to religious instruction that

resulted from the endeavour of earlier generations of Muslim activists to reach segments of society hitherto excluded from religious knowledge. What 'moral lessons' do the leaders of the Muslim women's groups dispense to their 'disciples'?

> Today, I will lecture about how a proper Muslim woman should fulfil her obligations of worship, [...] (and) how to show your devotion to realizing God's will in the here-and-now, [...] in your dealings with your in-laws and neighbours. The wives of the Prophet, peace be upon him, behaved as model wives and mothers...To emulate them is of utmost importance because life in the family today is shattered by many conflicts and disagreements. [...] We need to remind ourselves of the true meaning of Islam, of what God expects us to do. Nowadays money is on everyone's mind. Nowadays, some women want to have a greater say in family matters because they carry a greater burden. Our Prophet, peace be upon him, he said: only if you believe in God, if you follow the true teachings of Islam, will you be saved. Therefore we have to strive everyday to become a better Muslim, and teach ourselves the true meanings of Islam so that we can show others the path to God [...].
>
> (Aissetou, leader of a Muslim women's group
> in Bamako, Missira, August 2000)

While group leaders understand their teachings as an instruction in matters of traditional Islamic prescriptions and ritual practice, their lectures frequently reveal a tension between the traditional gender morals the leaders propagate and their emphasis on women's responsibility for collective moral renewal. This tension reflects on the dilemmas many urban women face in the wake of limited income opportunities, a greater financial responsibility for family subsistence and a persistent patriarchal gender ideology. Aissetou's 'moral lesson' illustrates that female leaders relate their admonishment that women should invite others 'to embark on the path to God' (*ka alasira ta*) to an emphasis on their own proper conduct. Disciples should show their dedication in proper ritual performance and in the cultivation of emotional capabilities essential to socially responsible conduct, among them *maloya* (modesty, 'shame'), *sabati* (endurance, patience), and a capacity for self-control and submissiveness (*munyu*) towards husbands and seniors. This view of emotional capabilities as elements of personal ethical reform bears resonances with Salafi-Sunni inspired thought articulated in female revivalist circles in Cairo (Mahmood, 2005). Still, in contrast to female *da'wa* and personal reform in Cairo, where women do not seem to view their public profession of their faith as essential to their mission, Muslim women in Mali emphasize the public and collective significance of their daily practices (Schulz, 2008a). Their public enactment of their newly found faith, as a means to convince others of the necessity to join their moral movement, transpires in a range of activities and practices in public and semi-public settings (Schulz, 2007, 2008a). Among these practices is the choice of 'decent' apparel, associated, by members of the movement of reform and by outsiders, with 'Islamic' ethics; and the performance of worship and other ritual obligations in spheres of greater public visibility.

These activities blur commonly accepted demarcations between public and domestic settings, and between gender-specific realms of practice. The particular objectives that Muslim women formulate reflect on the Western schooling paradigm with its emphasis on literacy and on the importance of 'building one's own viewpoint'. They are also the result of influences by transnational Muslim reformist trends and *da'wa* efforts; by institutions and discourses of a developmental state and of Western aid agencies; and finally, by local debates over proper Muslimhood polarized between representatives of a (heterogeneous) religious establishment and younger Muslim critics with new credentials of leadership. Because they recombine these influences in different ways, individual groups, teachers and leaders occupy divergent positions within the landscape of Muslim activism. The 'movement' of Islamic moral renewal in Mali therefore does not refer to a homogenous project. Constructions of the 'true' Muslim believer are effected through controversies among supporters of the movement, and between them and 'other' Muslims (Schulz, 2008b).

The emphasis many Muslim women place on the public declaration of their faith implies a departure from the prior relegation of female religiosity and devotional practices to an intimate, secluded space within the domestic realm. Whereas before, women's spiritual experiences were predicated upon their withdrawal from the area of worldly matters and mundane daily activities, Muslim women's public worship establishes a direct link between the performance of religiosity and its public profession. Muslim women's felt obligation to invite others to join their spiritual quest and to articulate their convictions in a national political arena runs against their own ideal of achieving spiritual accomplishment in relative seclusion. It complicates their claims that female propriety and a perfected religiosity should be achieved through acts that engage the individual believer in a personal relationship to God, far away from, and in disregard of, public scrutiny.

Muslim associations and the landscape of politico-religious patronage

The institutional basis of the Malian moral renewal movement is characterized by a diversity of organizational structures that, rather than fitting a neat classificatory scheme opposing 'formal' organizations to informal groups (as some accounts of civil society organizations in Africa and the Middle East would have it), display various degrees of institutionalization. Religious associations in Mali draw on conventions of religious mobilization that, in some areas of Muslim West Africa more than in others, arose out of the encounter with the colonial 'civilizing' mission. They are in line with a religious sociality associated with West African mystical traditions of Islam, and equally with the groups initiated by 'reform-minded' Muslims who, starting in the 1930s, challenged the hegemony of established religious clans. Some of them mobilize not only an urban following, but foster patronage ties between urban benefactors and rural hinterlands. These patronage networks offer services and infrastructure in the realm of schooling and religious education that the state fails to provide (Schulz, 2004). As such, Muslim associations in Mali sustain practices based historically on the intertwining of religious and political leadership in West Africa. As Villalon (1995) illustrates for Senegal, for instance, a characteristic feature of the Muride Sufi order, and a principal motor for its astounding success since French colonial rule, was its capacity to bind religious leaders and disciples into an 'economy of affection' based on reciprocal moral and material obligations.[12] Hence, the close interlocking of religious and political-strategic concerns in contemporary Mali should not be seen as a sign of a recent 'politicization' of religion, nor, for that matter, as an indication for the 'emptying out' of true religious conviction and doctrines by the rationale of modern state power. Contemporary Malian Muslim associations also echo forms of grass-root mobilization. These are currently mushrooming throughout the wider Muslim world, where these groups variously address women, disenfranchised or privileged segments of the urban youth (e.g. Le Blanc, 1998; Miran, 1998; Schulz & Janson, 2008; but see also Hock, 1998, chapter 5). In Mali, mobilization has been most successful among married women.

Until the 1970s, organized and regular gatherings of Muslim women were a rare phenomenon, limited to a few towns located in the southern triangle of Mali.[13] The urban, neighbourhood-based associations of 'Muslim women' have been largely a result of developments since the early 1980s. Many of them owe their existence to the initiative and infrastructural support of women whose employment in the administration and political structures under the former president Moussa Traore often secures them great prestige and influence at neighbourhood level. Some neighbourhood groups facilitate their leaders' endeavour to gain public recognition as the *présidente* of an (officially acknowledged) 'association' and thus as a representative of civil society. These forms of Muslim grass-roots mobilization should therefore not be seen as instances of 'primordial' social structures; nor do they signify a return to a former religious order. Rather, they are the result of recent reconfigurations of social and political alliances, and, as we will see below, also of the influences of Western donor-supported 'development' structures and discourses.

As mentioned earlier in this chapter, one important difference between these groups and their historical predecessors is that their social basis and age composition has changed. Nowadays, many participants of Muslim neighbourhood groups and networks come from the urban middle and lower-middle classes and represent a younger generation of married women. The organizational structures of the groups and the objectives of their gatherings have partly changed, too. As structures of sociability and financial support, Muslim women's groups often enjoy greater credibility than the credit-savings associations (French, *tontines*) that emerged in the wake of Structural Adjustment Policy. The terminology and organizational structure of these Muslim women's groups, including the roles of *présidente* and *trésorière*, indicates the extent to which they draw inspiration from Western donor-funded grass-roots organizations.

Most importantly, all these groups cut across social status categories and distinctions based on regional or ethnic origin. This explicit denial of social difference distinguishes these contemporary women's groups from the few historical predecessors in this area of the French Sudan, and also from the currently existing, conventional types of religious association in neighbouring Muslim majority countries.[14] The groups are also distinctive in the kinds of ties their leaders maintain to Malian patrons or foreign sponsors. As we will see, their immersion in both national and transnational networks of *clientage* reproduce differences of socio-economic status and religious prestige between leaders and followers, and also among different Muslim women's associations. These trajectories of internal differentiation will be explored below.

'Our submission to God's will makes us equal before Him': the dialectics of equality and hierarchy

Muslim women emphasize time and again both the joint nature of their search for closeness to God, and the fact that this common concern renders irrelevant any kind of status or other form of socio-economic difference between them. Yet the explicit dismissal of social distinction that both leaders and ordinary members of Muslim women's associations in Mali articulate does not foreclose the actual performance and reproduction of status difference. Social hierarchies, especially between ordinary members and group leaders, are established and reconfirmed in daily interaction, in both symbolic and material forms (Schulz, 2007). These patterns of inequality are reproduced, and sometimes reinforced, through the insertion of these women's groups into networks of funding and political patronage that operate at national, local and transnational levels. In other words, the hierarchies of status, wealth and prestige *among* Muslim women and their logic of reproduction, is predicated importantly on the differential insertion by ordinary group members and by their leaders into a network of national and transnational religious patronage and sponsoring. There thus inheres a strong tension in the existence and workings of Muslim women's groups. Patronage affiliation and a hierarchy of religious prestige are constitutive features of these groups, yet this hierarchical principle simultaneously undercuts precisely the basis of equality and shared existential background to which these women have recurrently laid claim.

As mentioned earlier, the high tides of foreign-sponsored Islamic welfare infrastructure and educational projects have been over for more than a decade.[15] Whereas the reasons for this development are complex, histories of mismanagement of funds by local recipients of these donations certainly played a role in the drying up of funds. Some long-distance sponsoring provided primarily by individual benefactors from Saudi Arabia, Egypt, Libya, Iran and countries of the Maghreb, is still effective.[16] Whatever the background for the decline in international funding is, it is a fact that even in the heyday of foreign-origin *da'wa* intervention in Mali, the lion's share of institutional support was directed to and administered by male recipients. This pattern has not changed. Moreover, as several female group leaders complained, funding for their groups has become more erratic and less predictable.[17] Only a few initiators of women's groups were willing to talk about the funding structure of their groups, but those who did explained that they received, at least at the initial phase, some kind of funding. Their international and national sponsors seemed to be, without exception, men. Yet, it is usually the initiators and leaders of the Muslim women's groups who act as the recipients of financial contributions in the name of group members, thereby facilitating a chain of sponsoring that connects different levels and forms of intervention.[18] Even if the benefactor is a locally renowned personality, rare are the cases in which ordinary group members may approach him directly.[19] Muslim women's group leaders, then, due to their intermediary positions between ordinary group members and sponsors at the national and international level, constitute and negotiate the crucial nexus point of extraversion between transnational influences and local members' aspirations and hopes.[20]

Group leaders usually access networks that connect to sponsors in the Arab-speaking world through their own male relatives and in-laws. The latter may maintain trade connections to other areas of the Muslim world or, having themselves spent years abroad, may even draw on long-standing forms of co-operation with business partners in the countries in question. There are also some group leaders who cannot rely on kin-mediated direct access to foreign sponsors. Nevertheless, many of them capitalize on their considerable reputation at the neighbourhood level to solicit some wealthy patron – very often someone of merchant origin who happens to live in the same neighbourhood or to whom the *présidente* happens to have a close contact based on long-standing acquaintance, remote matrimonial ties or on a common regional origin. This suggests that in many cases, a prerequisite for a women leader's access local and international sponsors is her capacity to mobilize fellow Muslim women, a capacity that is stands in mutually constitutive relationship with her moral reputation.

Notably, the relations to local or international sponsors the *présidentes* cultivate usually does not entitle them to any regular contribution. Rather, these women's attempts to position themselves in these local and transnational ties of sponsoring should be seen as attempts to tap into networks that *may* prove helpful in gaining institutional support or some financial input in case of need. The fact that the rewards of these women's ties to sponsors are so unpredictable and precarious is significant in several respects. Not only does this precariousness constitute an important dimension of practices of extraversion that may apply to other contexts and religious networks, it is also something which so far has not received sustained empirical research. The unpredictable outcome of the *présidentes'* clientilist affiliations to wealthy sponsors is also significant because it helps reassess the validity of frequent rumours, entertained mostly by critics of Islamic moral renewal, that Muslim women's leaders use their brokering position mostly for their own benefit. Judging from my close acquaintance with some of these leaders, I submit that such allegations have little empirical grounding. Only in few cases were female protagonists of Islamic renewal in a position to capitalize on their ties to foreign and local sponsors to receive grants for a study at institutions of higher religious learning in North Africa. Yet none of these women were *présidentes* because their younger age and unmarried status does not allow them to occupy a formal leadership position. Their access to grants was mediated through personal connections, mostly through the ties their families had to the respective educational institutions in the Maghreb. All of this suggests that the *présidentes'* positions within national and transnational networks of patronage, sponsoring and learning, and their attendant practices of extraversion, are only one among several factors that reproduce differences within the field of Islamic moral renewal, and among Muslim women in particular.

Conclusion

This chapter examined female Muslim activism not in terms of the threat it allegedly poses to the secular nation state, but by stressing the role Muslim women's religious sociality plays in making and remaking society from within. With its emphasis on the socially constitutive effects of Muslim women's activities, the chapter sought to engage from a different perspective current debates about political Islam, its new public faces and purported dangers, and, more generally, about the seemingly novel roles and aspirations that religious movements claim in contemporary African politics. By situating the current movement historically, it sought to convey the social import of this activism; that is, its contemporary capacity to appeal to and articulate the moral apprehensions of a broad range of followers. As structures of sociability, learning and mutual financial support, women's 'religious' associations allow them to claim collective relevance at the interstices of domestic, semi-public and public settings, and to initiate new nodes of articulation between society and the state. As such, female Muslim activism in Mali can be seen as a particular modality of 'politics from below', one that aims not at political protest, but at the transformation of the personal and the social.

To qualify these findings, two cautionary remarks are relevant. First, the insertion of protagonists of Islamic renewal into a landscape of transnational ties, intellectual influences and symbolic references has paradoxical consequences. Its transnational orientation and locations give rise to new conditions of intervention in the national arena, yet simultaneously contributes to the reproduction of two trajectories of inequality. One is the unequal position of male and female initiators within the moral reform movement, which is as much a precondition as it is a consequence of women's and men's differential insertion into local and transnational religious patronage networks. The second is the reinforcement of the differential social and economic standing of leaders and ordinary group members, a rather closed universe of social hierarchy from which only few (and usually already privileged) supporters of the movement may escape.

Secondly, while this chapter stressed the need to consider what material and political considerations go into the activities of supporters of Islamic moral renewal, this should not be mistaken for an explanation of the existence of Muslim women's groups by their material and political significance alone. It would be highly problematic to reduce the *raison d'être* of Muslim associations to their attempt to tap into national patronage structures and transnational networks of religious sponsoring, or to dismiss their endeavour as being guided by short-term materialist considerations. These interpretations not only ignore the historical close intertwining of notions of spiritual and material salvation (Schulz, 2006), but also fail to explain why the endeavour to gain knowledge in religious matters has become a key metaphor of Muslim women's activities and mobilization. To sketch a nuanced picture of 'religious' practices of extraversion, one that goes beyond the economic meanings of the concept as it was originally conceived by Samir Amin, we need to make room for the possibility that practices of extraversion may aim at more than immediate material advantage. Muslim women's educational endeavour offers an entry point to expand our understanding of the aims at which Muslim women's acts of 'extraversion' – no longer understood as practices geared exclusively towards material benefit – are directed. Muslim women's eagerness to civilize themselves through the acquisition of knowledge relevant to their ethical quest is in line with a long-standing agenda of Muslim reformism since late colonial rule. But their learning activities are also novel inasmuch as they reflect a new sensibility and understanding of religious subjectivity. Religious knowledge has left the restricted realm of schools and moved into a more accessible arena of generalized learning. Simultaneously, the significance of religious education is changing. 'Learning' comes to embody the promises of ethical self-improvement, of a more immediate access to the written foundations of Islam, and thus the possibility to live one's faith without relying on the intercession of religious experts. Thus, as much as Muslim women emphasize a sense of sharing and of collective responsibility, their institutions of religious sociality also support them in their simultaneous individualizing and civilizing endeavour of personal self-making.

Notes

1. Unless noted otherwise, all foreign terms are rendered in Bamanakan, the lingua franca of southern Mali.
2. Notable exceptions are Weiss (2002) and Kaag (2007) who record how international and national Islamic NGOs provide social services in sectors from which the state has withdrawn in Ghana and Chad (see Ghandour, 2002).
3. For example, Brenner (1993b), Otayek (1993), Gomez-Perez (1994, 2005), Glew (1996), Loimeier (1997), Miran (1998, 2005), Renders (2002), Weiss (2002), Kane (1997), Soares (2005).
4. For example, Dale Eickelman and Jon Anderson (1999), Launay and Soares (1999), Salvatore (1999), Hirschkind (2001), Schulz (2003), Meyer and Moors (2006).
5. See Fraser (1992), Meehan (1995). But see Postone (1992).
6. These renewal efforts did not start with colonialism but dated back to earlier reforms initiated by Muslims with affinities to the mystical traditions of Islam and attendant Sufi orders (Arabic, *turuq*, see Loimeier, 2003).
7. In Mali, Guinea and Northern Ivory Coast, Muslim reformists' efforts materialized in a stricter dress code, the denouncement of certain lifecycle rituals, and, in some groups, the adoption of a prayer posture associated with Wahhabi doctrine (Kaba, 1974; Amselle, 1985; Triaud, 1986; Launay, 1992; Masquelier, 1999).
8. Association Malienne pour l'Unite et le Progres de l'Islam.
9. Herein he differed from his predecessor, President Keita (1960–1968), under whose single-party rule manifestations and actors of Islam had been widely sidestepped.
10. Members came from diverse socio-economic background. Few were married to men whose employment in the state administration secured them a steady income. Most women worked in sectors of the informal economy with low profit margins. About 30 per cent of the group members were de facto single household heads because their husbands were either handicapped or unemployed.
11. In the early days of her group attendance, Fanta wore dressy robes, a somewhat peculiar form of attire for this kind of setting.
12. Also see Amselle (1977), Kaba (1974), Launay (1992).
13. Compared to colonial Senegal and Nigeria, where urban Muslim associations geared towards religious education became important urban institutions in the 1920s (e.g. Villalon, 1995, chapter 5; Reichmuth, 1996; Halidou, 2005), Muslim associations in Mali merged at a slower pace. Most associations were created in Bamako and Kankan which became the newly emerging centres of

colonial French Sudan and the key sites of Muslim reformist activities (Kaba, 1974; Brenner, 2001).

14. In Senegal, the latter type of association is generally tied to Sufi social organization and based on notions of human intercession and hierarchies of religio-genealogical prestige (Villalon, 1995, chapter 4; Evers Rosander, 1997, 1998). In Muslim women's neighbourhood groups in Mali, in contrast, relations to their leaders are structured by a range of sources and notions of authority (Schulz, forthcoming).

15. Financial support was provided by Islamic NGOs and by individual sponsors in Saudi Arabia, Libya and other countries of the Arab world whom the Malian recipients had met during their foreign travels and business operations.

16. There are indications that some Islamic NGOs have recently intensified their activities in West Africa (Kaag, 2007; see Salih, 2002). I did not find empirical evidence for this interpretation.

17. These leaders were without exception women with whom I had a longer and confidential relationship.

18. For reasons of etiquette and respect, ordinary group members cannot control the *présidente*'s handling of financial matters. This lack of accountability occasionally prompts group members to gossip about their leader's (ab)use of their funds.

19. As we will see later, 'ordinary' Muslim women tend to rely on the mutual financial support that group membership allows for, rather than on direct support from their *tontigi*.

20. Leaders and group members refer to these sponsors as 'hope' (*jigi*), a term that illustrates that they operate within the moral universe of patronage typical of other social security networks in Malian society.

References

Amselle, Jean-L. (1977), *Les Négociants de la Savane*, Paris: Éditions anthropos.

Amselle, Jean-L. (1985), 'Le Wahhabisme à Bamako (1945–1985)' *Canadian Journal of African Studies*, 19 (2), pp. 345–57.

Augis, E. (2002), *Dakar's Sunnite Women: The Politics of Person*, Ph.D. Thesis, Department of Sociology, University of Chicago.

Baker, Houston (1994), 'Critical memory and the black public sphere' *Public Culture* 7 (1), pp. 3–33.

Bayart, J. F. (ed.) (1993), Religion et modernité politique en Afrique noire: Dieu pour tous et chacun pour soi. Paris: Karthala.

—— (2000), 'Africa in the world: A history of extraversion' *African Affairs*, 99, pp. 217–67.

Bornstein, E. (2003), The Spirit of Development: Protestant NGOs, Morality, and Economics in Zimbabwe. London, New York: Routledge.

Brenner, L. (1993a), 'Constructing Muslim identities in Mali', in L. Brenner (ed.), *Muslim Identity and Social Change in Subsaharan Africa*, Bloomington: Indiana University Press, pp. 59–78.

Brenner, L. (1993b), 'La culture arabo-islamique au Mali' in Otayek, R. (ed.), *Le radicalisme islamique au sud du Sahara*, Paris, Talence: Karthala, pp. 161–195.

Brenner, L. (2001), *Controlling Knowledge. Religion, Power and Schooling in a West African Muslim Society*, Bloomington and Indianapolis: Indiana University Press.

Brönig, M. & Weiss, H. (eds) (2006), *Politischer Islam in Westafrika*, Eine Bestandsaufnahme, Münster: Lit Verlag.

Calhoun, Craig (ed.) (1992), *Habermas and the Public Sphere*, Cambridge, MA: M.I.T Press.

De Jorio, R. (2009), 'Between dialogue and contestation: Gender, Islam, and the challenges of a Malian public sphere' *Journal of the Royal Anthropological Institute*, 15 (1), pp. 95–111.

Eickelman, D. & Anderson, J. (1999) 'Redefining Muslim publics' in Eickelman, D., Anderson, J. (eds), *New Media in the Muslim World*, Bloomington, Indianapolis: Indiana University Press, pp. 1–18.

Eickelman, D. & Salvatore, A. (eds) (2004), *Public Islam and the Common Good*, Leiden: Brill. Eisenstadt, S.N.

Englund, H. (2003), 'Christian independency and global membership: Pentecostal extraversions in Malawi', *Journal of Religion in Africa*, 33 (1), pp. 83–111.

Evers Rosander, E. (1997), 'Le dahira de Mam Diarra Bousso de Mbacké' in Evers Rosander, E. (ed.), *Transforming Female Identities: Women's Organizational Forms in West Africa*, Uppsala: Nordiska Afrikainstitutet, pp. 160–174.

Evers Rosander, E. (1998), 'Women and Muridism in Senegal: The Case of the Mam Diarra Bousso Daira in Mbacké' in Ask, K. T. & Marit, T. (eds), *Women and Islamization. Contemporary Dimensions of Discourse on Gender Relations*, Oxford, New York: Berg, pp. 147–176.

Fourchard, L., Mary, A. & Otayek, R. (eds) (2005), *Entreprises religieuses transnationales en Afrique de l'Ouest*, Paris and Ibadan: IFRA, Karthala.

Fraser, N. (1992), 'Rethinking the public sphere: A contribution to the critique of actually existing democracy' in Calhoun, C. (ed.), *Habermas and the Public Sphere*, Cambridge, MA: MIT Press, pp. 109–142.

Ghandour, Abdel-R. (2002), *Jihad humanitaire: enquête sur les ONGs islamiques*, Paris: Flammarion.

Gifford, P. (ed.) (1995), *The Christian Churches and the Democratisation of Africa*, Leiden: Brill.

Gifford, P. (1998), *African Christianity. Its Public Role*, Bloomington: Indiana University Press.

Glew, R. (1996), 'Islamic associations in Niger' *Islam et Société au Sud du Sahara*, 10, pp. 187–204.

Gomez Perez, M. (1994), 'L'islamisme à Dakar: d'un contrôle social total à une culture du pouvoir' *Afrika Spektrum*, 1, pp. 79–98.

Gomez Perez, M. (2005), 'Généalogies de l'islam reformiste au Sénégal: figures, savoirs et réseaux' in Fourchard, L., Mary, A. & Otayek, R. (eds), *Entreprises religieuses transnationales en Afrique de l'Ouest*. Ibadan, Paris: IFRA, Karthala, pp. 43–72.

Halidou, O. (2005), *Engaging Modernity. Muslim Women and the Politics of Agency in Postcolonial Niger*, Madison, WI: The University of Wisconsin Press.

Hirschkind, C. (2001), 'Civic virtue and religious reason: An Islamic counterpublic' *Cultural Anthropology*, 16 (1), pp. 3–34.

Hock, C. (1998), 'Muslimische Reform und staatliche Autorität in der Republik Mali seit 1960', Ph.D. Thesis, University of Bayreuth.

Hofer, Katharina (2003), 'The Role of Evangelical NGOs in International Development: A Comparative Study of Kenya and Uganda' *Africa Spectrum* 38(3), pp. 375–398.

Joseph, S. (1997), 'Gender and Civil Society (interview with Joe Stork)' in Beinin, J. J. S. (ed.), *Political Islam, Essays from Middle East Report*, London, New York: I.B. Tauris Publishers, pp. 64–82.

Kaag, M. (2007), 'Transnational islamic NGOs in Chad: Islamic solidarity in the age of Neoliberalism' *Africa Today*, 54 (3), pp. 3–18.

Kaba, L. (1974), *The Wahhabiyya: Islamic Reform and Politics in French West Africa*, Evanston, IL: Northwestern University Press.

—— (2000), IslamLevtzion, N. & Powell, R. (eds), *The History of Islam in Africa*, Athens, Oxford, Cape Town: Ohio University Press, James Currey, David Philip, pp. 189–208.

Kane, O. (1997), 'Muslim Missionaries and African States' in Rudolph, S. & Piscatori, J. (eds), *Transnational Religion and Fading States*, Boulder, CO: Westview Press, pp. 47–62.

Kane, O. & Triaud, J. L. (1998), 'Introduction' in Kane, O. & Triaud J. L. (eds), *Islam et islamismes au sud du Sahara*, Aix-en-Provence, Paris: Irenam, Karthala, pp. 5–30.

Kleiner-Bosaller, A. & Loimeier, R. (1994), 'Radical Muslim Women and Male Politics in Nigeria' in Reh, M. & Ludwar-Ene, G. (eds), *Gender and Identity in Africa*, Münster and Hamburg: LIT Verlag, pp. 61–69.

Launay, R. (1992), *Beyond the Stream. Islam and Society in a West African Town*, Berkeley, Los Angeles, Oxford: University of California Press.

Launay, R. & Soares, B. (1999), 'The formation of an "Islamic sphere" in French Colonial West Africa' *Economy and Society*, 28 (3), pp. 467–89.

Le Blanc, M. N. (1998), 'Youth, Islam and changing identities in Bouaké, Côte d'Ivoire', Ph.D. Thesis, University of London.

(1999) 'The production of Islamic identities through knowledge claims in Bouaké, Côte d'Ivoire' *African Affairs*, 98 (393), pp. 485–509.

Loimeier, R. (1997), *Islamic Reform and Political Change in Northern Nigeria*, Evanston, IL: Northwestern University Press.

Loimeier, R. (2003), 'Patterns and peculiarities of Islamic reform in Africa' *Journal of Religion in Africa*, 33 (3), pp. 237–62.

Mahmood, S. (2005), *Politics of Piety. The Islamic Revival and the Feminist Subject*, Princeton, Oxford: Princeton University Press.

Marshall-Fratani, R. (1998), 'Mediating the Global and the Local in Nigerian Pentecostalism' *Journal of Religion in Africa*, 28 (3), pp. 27–315.

Masquelier, A. (1999), 'Debating Muslims, Disputed Practices: Struggles for the Realization of an Alternative Moral Order in Niger in Comaroff, J. & Comaroff, J. L. (eds), *Civil Society and the Political Imagination in Africa. Critical Perspectives*, Chicago, London: The University of Chicago Press, pp. 218–250

Meehan, J. (1995) (eds), *Feminists Read Habermas: Gendering the Subject of Discourse*, New York, London: Routledge.

Meyer, B. (2004), ' "Praise the Lord": Popular cinema and pentecostalite style in Ghana's new public sphere' *American Ethnologist*, 31 (1), pp. 92–110.

Meyer, B. & Moors, A. (eds) (2006), *Religion, Media, and the Public Sphere*, Bloomington: Indiana University Press.

Miran, M. (1998), 'Le Wahhabisme à Abidjan: dynamisme urbain d'un islam reformiste en Côte d'Ivoire contemporaine (1960–1996)' *Islam et Sociétés au Sud du Sahara*, 12, pp. 5–74.

Miran, M. (2005), 'D'Abidjan à Porto Novo: associations islamiques et culture religieuse réformiste sur la Côte de Guinée' in Fourchard, L. A. M. R. O. (ed.), *Entreprises religieuses transnationales en Afrique de l'Ouest*, Ibadan and Paris: IFRA, Karthala, pp. 43–72.

Otayek, R. (ed.) (1993), *Le radicalisme islamique au sud du Sahara*, Paris, Talence: Éditions Karthala.

Postone, M. (1992), 'Political Theory and Historical Analysis' in Calhoun, C. (ed.), *Habermas and the Public Sphere*, Cambridge, MA; London, UK: MIT Press, pp. 164–180.

Ranger, T. (2003), 'Evangelical Christianity and democracy in Africa: A continental comparison' *Journal of Religion in Africa*, 33 (1), pp. 112–17.

Reichmuth, S. (1996), 'Education and the growth of religious associations among Yoruba Muslims – the Ansar-ud-Deen society in Nigeria' *Journal of Religion in Africa*, 26 (4), pp. 364–405.

Renders, M. (2002), 'An ambiguous adventure: Muslim organizations and the discourse of "Development" in Senegal' *Journal of Religion in Africa*, 32 (1), pp. 61–82.

Salih, M. (2002), 'The Promise and Peril of Islamic Voluntarism', Occasional Paper, Center for African Studies, University of Copenhagen.

Salvatore, A. (1999), 'Global influences and discontinuities in a religious tradition: Public Islam and the "New" Sari'a' in Füllberg- Stolberg, K., Heidrich, P. & Schöne, E. (eds), *Dissociation and Appropriation. Responses to Globalization in Asia and Africa*, Berlin: Das Arabische Buch, pp. 211–234.

Sanankoua, B. (1991), Les associations féminines musulmanes à Bamako' in Sanankoua, B. L. B. (ed.), *L'enseignement islamique au Mali*, Bamako: Editions Jamana, pp. 105–126.

Schulz, D. (2003), 'Charisma and brotherhood revisited: Mass-mediated forms of spirituality in Urban Mali' *Journal of Religion in Africa*, 33 (2), pp. 146–71.

Schulz, D. (2004), 'Islamic revival, mass-mediated religiosity and the moral negotiation of gender relations in urban Mali', Habilitation Thesis, Free University of Berlin.

Schulz, D. (2006a), 'Promises of (Im)mediate salvation. Islam, broadcast media, and the remaking of religious experience in Mali' *American Ethnologist*, 33 (2), pp. 210–29.

——— Schulz, D. (2006b), Morality, Community, "Public-ness": shifting terms of debate in the Malian Public. *In* Meyer, B. A. M. (ed.), *Religion, Media, and the Public Sphere*, Bloomington: Indiana University Press, pp. 132–151.

——— Schulz, D. (2007), Competing sartorial assertions of femininity and Muslim identity in Mali. Fashion Theory 11(2/3): 253–280.

Schulz, D. (2008a), 'Piety's manifold embodiments: Muslim women's quest for moral renewal in urban Mali' *Journal for Islamic Studies*, 28, pp. 26–93, Special Issue on "Reconfigurations of Gender Relations in Africa", Janson, M. & D. Schulz, guest editors.

Schulz, D. (2008b), '(Re)Turning to proper Muslim practice: Islamic moral renewal and women's conflicting constructions of Sunni identity in urban Mali' *Africa Today*, 54 (4), pp. 21–43.

Schulz, D. (2010) 'May God Let me Share Paradise with My Fellow Believers'. Islam's "Female Face" and the Politics of Religious Devotion in Mali' in Buggenhagen, B. A., Jackson, S. & Makhulu, A. M. (eds), *Hard Times, Hard Work. Perspectives on the Politics of Agency in Africa*, Berkeley: University of California International and Area Studies Digital Collection.

Schulz, D. & Janson M. (2008), 'Introduction. Reconfigurations of gender relations in Africa' *Journal for Islamic Studies*, 28, pp. 1–12, Special Issue on 'Reconfigurations of Gender Relations in Africa', Janson, M. & Schulz, D. guest editors.

Soares, B. (2005), *Islam and the Prayer Economy. History and Authority in a Malian Town*, Ann Arbor: University of Michigan Press.

——— Soares, B. (2006), 'Islam in Mali in the Neoliberal Age' *African Affairs*, 105, pp. 77–95.

Soares, B. & Otayek, R. (eds) (2007), *Islam and Muslim Politics in Africa*, New York: Palgrave Macmillan.

Triaud, J.L. (1986), 'Abd al-Rahman l'Africain (1908–1957), pionnier et precurseur du wahhabisme au Mali' in Carré, R. O. & Dumont, P. (eds), *Radicalismes Islamiques*, vol. 2, Paris: Harmattan.

Villalon, L. (1995), *Islamic Society and State Power in Senegal: Disciples and Citizens in Fatick*, Cambridge: Cambridge University Press.

Weiss, H. (2002), 'Reorganising social welfare among Muslims: Islamic voluntarism and other forms of communal support in Northern Ghana' *Journal of Religion in Africa*, 32 (1), pp. 83–109.

Part II

'Religion Between State and Society'

5
Da`wa and Politics in West Africa: Muslim *Jama`at* and Non-Governmental Organizations in Ghana, Sierre Leone and The Gambia

David E. Skinner

In this chapter I examine the interplay of Islam, politics and development in three West African states during the postcolonial era. The focus is on the creation and maintenance of Islamic space and efforts by Muslims to expand their political, economic and social influence in these states through the formation of Non-Governmental Organizations (NGOs) and their interaction with governments and international agencies. The Muslim NGOs may operate on four levels: local, regional, national and international. However, many, perhaps most, are interested in developing their local resources and influence and may not be involved in national development. All face the reality of operating within the national and international systems in order to gain material support and recognition of their goals but the question remains how their activities contribute to development of the social, economic and political institutions of their national states. Two critical problems in this question are the lack of coordination of the separate activities and goals of Muslim NGOs and the tendency for NGOs to avoid consultation with national governmental agencies when they pursue their individual objectives. A third, and related, problem has been the difficulty of all three national states to create stable governmental systems and to provide adequate national development programmes during the post-colonial era.

In Africa, Islam was introduced by means of long-distance trade and the transmission of knowledge through educational institutions. The missionary aspect of Islam is based on the concepts of *da`wa* (calling/ mission) and *jihad* (exertion/struggle) which encourage Muslims to preach and promote the fundamentals of their faith and to behave as role models for humanity. *Jihad* as 'holy war' is meant to be used under conditions of violent opposition and attack against the *ummah* (the Muslim community) by unbelievers or hypocrites. The preferred means of spreading Islam to those who are ignorant is through preaching and teaching about the faith by knowledgeable Muslims who may be pursuing any occupation but who have the skill and opportunity to introduce others to the true path. Although Islam was introduced to North Africa through conquest by Arab forces, who then established political hegemony, generally Islam spread throughout Africa by means of the skilled educational qualities of merchants and religious specialists who were involved in long-distance trade and education networks.

Islam in West Africa

Because of the long-distance trade networks dominated by Muslim merchants and families of scholars, West Africa, particularly in the Sahel region, in large interior market towns and in some Atlantic coast ports, was a fruitful area for the establishment of Islam. Over several centuries Islam became associated with important empires and kingdoms, and many kinship groups provided skilful leadership in spreading the faith along the long-distance trade routes. Particular kinship groups became notable for their knowledge of and zeal in the propagation of Islam. Religious specialists from these families, especially from the Mande, Fula or Hausa ethnic groups, were active in many areas of western Africa and challenged the French and British for authority during the colonial period. Through their missionary programmes, they induced members of other ethnic groups to convert to and become missionaries for Islam. These Muslim notables were well-placed to assume important leadership roles in postcolonial West Africa.[1]

This chapter focuses on the development of Islamic education and political activities in three West African nation-states in the postcolonial era. The Gambia, Ghana and Sierra Leone were selected for comparative study because each was subject to British colonial administration, yet each has a distinctive composition of ethnic groups and Muslim/non-Muslim populations. One objective of this study is to evaluate the different intensities of Islamic activity in the three nation-states and to decide whether Muslim leaders play more distinctive roles in a country where the percentage of Muslims in a nation-state is high rather than relatively low.[2]

Although The Gambia, Ghana and Sierra Leone experienced varying degrees of Islamic influence and differing rates of conversion, by the end of the nineteenth century each of them had well-established systems of education headed by professional clerics whose families had extensive contacts with Muslim-dominated networks. By using the already established educational base, in spite of British programmes to promote Christianity, Islam had made significant progress in all three by the end of the colonial period. Islamic education programmes had been introduced which were beginning to compete with Christian and colonial schools, and modern Islamic organizations were converting many residents to the faith. In The Gambia a group of men formed the Young Muslim Society in 1936, the same year that the Ga Aborigines Muslim Mission was founded in the Gold Coast (Ghana), while the Sierra Leone Muslim Congress may have been formed as early as 1928.[3] These and other organizations were responsible for creating a basis for the Islamic revitalization which has occurred in the postcolonial period.

Since the 1950s there has been an increase in and intensification of Islamic activities in these three nation-states. Muslim leaders have organized to fulfil three essential roles: first, to assure that Muslims understand and perform their responsibilities; secondly, to convert non-Muslims to Islam; and thirdly, to promote the political influence and economic wellbeing of their communities. To fulfil these goals they have founded Islamic organizations which engage in a variety of social, economic and political activities. The most significant of these activities include organizing the *hajj* requirements for prospective pilgrims each year, providing for the prayers and celebrations of the feast days, building new mosques and medical clinics, establishing educational institutions, recruiting teachers, raising funds to send scholars abroad for higher education, founding preachers' societies, and acting as liaison for their members with local and national governments and international agencies.

During the period 1960 to 1990 there was a remarkable increase in the number and variety of organizations which aimed to serve Islam. While the individual cleric with his small school still plays a vital role in the strengthening of Islam in many areas of West Africa, the modern trend has been the formation of large, centralized organizations which operate school systems, present conferences, print literature and send missionaries into the field. Certainly, the Ahmadiyya Muslim Mission has provided a model for this type of organization, but many new movements have been established during the postcolonial period. Only a few of the several hundred can be mentioned in this brief survey. The activities of these Muslim NGOs illustrate the roles they are attempting to play in the development of Islamic institutions and influence, and also their relationships with their national governments and with international agencies.

In The Gambia, the Islamic Union, the Gambia Muslim Association and the Islamic Solidarity Association of West Africa all operate schools and organize conferences.[4] For more than 30 years the Gambia Muslim Association has run the successful Muslim High School (opened in 1975) in Banjul. A relatively new organization, the Islamic Cultural Union of The Gambia, was founded in 1984. It is a branch of the same organization based in Dakar, Senegal. In The Gambia the founding patron was al-Hajj Abdullai Jobe, Imam of the Banjul central mosque. The Islamic Cultural Union sponsors conferences, distributes literature and holds training courses for missionaries. Every year it participates in a joint symposium with the Senegal branch. The purpose of these symposia is to publicize the activities of the great *murabitun* (teachers and scholars) of Senegal and The Gambia.[5]

Ghana, too, has a variety of organizations. The 'Anbariya Islamic Institute of Tamale', founded by al-Hajj Yusufu Ejura, reputedly has the largest school in West Africa, and it attracts students from many nation-states. The Ghana Muslim Mission, the Muslim Community and the Supreme Council for Islamic Affairs all support schools, public preaching and conferences. A novel feature of Muslim missionary work in Ghana, where Islam struggled to survive in a strong Christian environment, has been the building of new villages exclusively for Muslim converts.[6] A firmly Muslim subdivision of Accra, Nima, which was settled mainly by migrants to the area, has more than 20 schools affiliated to different organizations.[7] In the north, a traditional area of strength for Islam, there are hundreds of affiliated and unaffiliated schools.[8]

The situation is the same in Sierra Leone. Some of the organizations which specialize in education and missionary work are the Sierra Leone Muslim Brotherhood, the Kankaylay Muslim Mission (Sierra Leone Muslim Women's Educational Institute) which runs a women's vocational school and other schools, Ansarul Islamic Mission (65 primary schools, 12 secondary schools and two colleges in 1988) and the Sierra Leone Islamic Federation, among many others. The Sierra Leone Islamic Federation held a conference on '*al-hajj*: Its Rights and Rites' in Freetown on 21 July 1988.[9] An influential organization, the Supreme Islamic

Council, sponsored a symposium in memory of al-Hajj Jibril Sesay in Freetown on 14 October 1988. Shaikh Sesay was probably the most highly respected Muslim in the country; he had advanced degrees in law, theology and Arabic from al-Azhar University, had been a leading educator, a political activist, Imam of a large mosque, ambassador to Egypt and had recently be selected as the first *mufti* of Sierra Leone. The symposium was opened by the First Vice President of Sierra Leone, Hon. A. B. Kamara and other officials.[10]

In studying the various educational, missionary and service programmes of the many Islamic organizations in the three nation-states one finds a pattern of competition for resources and political influence. The Gambia, Ghana and Sierra Leone have been influenced by rather distinctive agents of Muslim missionary work and are at different stages of 'islamization', but they all share one common characteristic. There is a lack of Islamic unity among the hundreds of *jama`at* and Muslim non-governmental organizations (NGO) in all three countries. The divisions are based on ethnic identity or country of origin, traditional Muslim educational patterns versus modernization of school facilities and curriculum, sectarianism or different views of what constitutes orthodoxy, access to political power and economic resources, and relations with international agencies. In none of the three countries is there a truly national Muslim organization which unifies all Islamic associations or communities, although attempts have been made to form a single federation and, at times, such organizations have claimed to be the national representative of Muslims in the nation.

A large number of organizations compete for members, funds, political influence and international recognition. Muslim organizations simultaneously operate within local, national and international political arenas. As we have seen, they are engaged in self-improvement schemes and develop programmes to promote their institutions and their goal of *da`wa*. These schemes and programmes contribute to the infrastructure through the construction of schools, mosques, health clinics and housing. In the process of these endeavours the organizations produce jobs for builders and maintenance persons, teachers, administrators and other workers. In turn, they help to educate the subsequent generations of Muslims for positions in education, public service and community leadership. Because of their competitive programmes for development many of these organizations function as political entities as they try to enhance their influence relative to other organizations.[11] In some instances, this situation induces Muslim organizations to create foreign affairs divisions and to behave as if they were autonomous governments.

Diversity and dissension

There are many situations in which Muslims have definite political interests. Sometimes political activities remain internal to the Muslim organizations, while at other times the issues transcend internal affairs and involve the organizations in national or international political, social and economic relations. Even when organizations experience internal disputes the situation may become public and impinge on local, national or international politics. Furthermore, local, national and international politics have consequences for the internal functioning of Muslim organizations. None of the three nation-states has a Muslim government, although The Gambia and Sierra Leone are members of the Organisation of the Islamic Conference (OIC),[12] and none of them presently has a Muslim political party.[13] Muslim organizations have become adept at manoeuvring within both the secular framework of national politics and the Islamic context of international aid and affiliation.

Indeed, Muslim disunity has been intensified through political competition in the national arena and through economic competition in international affairs. In The Gambia organizations vie for support from overseas sources. The World Islamic Call Society of Tripoli, Libya (WICS), is very actively supporting the development of schools, centres, medical facilities and agricultural programmes throughout West Africa. There has been a strong Islamic Call Society movement in The Gambia since the 1980s, and several Gambian missionaries are partially funded by the Society which has its own Gambian organization.[14] Representatives of Gambian Muslim organizations helped to form the Islamic Call Coordinating Committee in September of 1988. This action followed a visit by a delegation of Gambians to Tripoli in August. Several proposals for funding were presented to the Minister for Islamic Affairs and to the Islamic Call Society.[15] The search for financial aid frequently leads to divisions among Muslims and the formation of rival organizations with similar names. For example, again in The Gambia, one finds the Islamic Solidarity Association of West Africa and the Islamic Solidarity Association of The Gambia, both clearly interested in seeking aid from the Islamic Solidarity Fund which is administered by the Organization of the Islamic Conference.

In Sierra Leone, political party politics played a crucial part in the formation of the Sierra Leone Islamic Federation in April 1984. Al-Hajj S. A. T. Koroma was the founder/president of the Supreme Islamic Council and he had been a principal leader of the All Peoples' Congress (APC) in its early days. He was Minister of Social Services in the government in 1984 but was relieved of his position and was subsequently given a vote of 'no confidence' by the Supreme Islamic Council executive committee. As a result, he lost his presidency and left the Council to form the Islamic Federation.[16] Why? Political party affiliation and a powerful position within the dominant party provided some Muslim leaders and their organizations with advantages. The foreign minister in the 1990 government of Sierra Leone was al-Hajj Abdul Karim Koroma, and his official duties frequently took him to Kuwait, Saudi Arabia, Libya and other important Muslim countries. Dr A. K. Koroma was also the president of the Sierra Leone Muslim Brotherhood and supervised the construction of a Muslim college in the town of Magburaka. Funding for this institution came primarily from Saudi Arabia.[17]

In Ghana, national politics played an important role in the formation and operation of Islamic organizations. Beginning with the demise of the Muslim Action Party in 1957, national governments, especially after 1966, intervened to create new organizations or to control the operations of existing ones. After the overthrow of Kwame Nkrumah's CPP in 1966 there was a succession of short-lived governments, all but two dominated by high-ranking military officers. Only with the government of the Peoples' National Defence Committee (PNDC), headed by Flight Lieutenant Jerry Rawlings, which came to power on 31 December 1981, was the pattern broken. The rapid turnover of governments and the distinctly minority position of Islam in Ghana greatly heightened the sensitivity of Muslim leaders to the need for strong internal political action and for creating solid ties with external Islamic governments and agencies. As a result, some Islamic organizations worked diligently to attach themselves to powerful political figures or dominant political parties and to attack rival organizations for being out of step with national priorities.[18]

Ethnic identity is an important cause of Islamic disunity in the three nation-states. There has been intense competition in The Gambia between the Wolof, who were prominent in Islamic affairs in Banjul and who were quicker to respond to British education, and the Mandinka, who were instrumental in spreading Islam in the Gambia River basin and who were more aloof from British influence. With respect to Islam, this division is reflected in the formation of the Gambia Muslim Association in March 1965 by elders of the Banjul Wolof community; its predecessors, the Young Muslim Society and the Young Renaissance Club were also organizations dominated by Wolof leaders. The Muslim Association started the Muslim High School and helped to improve the Banjul central mosque. The Banjul central mosque committee was almost exclusively composed of Wolof elders. This became an issue among non-Wolof Muslims as the ethnic composition of Banjul and nearby towns changed. The issue intensified when the Islamic Development Bank funded the construction of a new central mosque which some Muslims thought should become the national mosque of The Gambia. However, the committee of the old central mosque selected itself to be the committee for the new mosque. This decision was opposed by many who felt the new mosque represented all Gambians and who advocated a committee with a more balanced membership. Although the mosque was ready to be used in late 1988, its formal opening was delayed, in part because of this dispute.[19]

Ethnicity has long been the basis for the construction of mosques in Freetown, Sierra Leone. Ethnic division was supported by the British colonial administration when it recognized the existence of 'tribal' authorities in Freetown in the latter decades of the nineteenth century. A 'tribal' administrative system was officially legislated in 1905, by which time ethnic mosques had already appeared. This pattern persisted in Freetown where one found 'Limba mosques', 'Mandingo mosques', 'Temne mosques', 'Mende mosques', and so on. As a result there was no central or national mosque in Freetown.[20] In the 1990s, a mosque dispute in Freetown was indirectly related to ethnic divisions. A beautiful mosque had been under construction for several years. It was funded by the Lebanese Shi'a business community, and the designated Imam was Shaikh Houssein Chahade, a Shi'a Lebanese missionary with a degree from Qum in Iran. Except for perhaps a few dozen African Sierra Leoneans all the Shi'a in the country were Lebanese, and the great majority of Muslims were Sunni. As the mosque was designed to be quite grand and was centrally located in Freetown, many Sunni Muslims questioned the propriety of its construction as a Shi'a mosque.[21]

Ethnicity, or one might even say, nationality, is a sensitive issue in Ghana. Several centuries ago, Islam was introduced in the northern regions by Mande traders and clerics, and Islam was respected in some of the important kingdoms such as Dagomba and Gonja. On the other hand, in Ashanti and along the coast among the Ga and Fanti, for example, Islam was not present or was very weak until the colonial era. The British brought soldiers and police from Nigeria to assist in the administration of the Gold Coast, and many of them were Muslim Hausa, Fulani or Yoruba. Others came to trade and, later, to establish Muslim schools. They settled in subdivisions usually called *zongo*. The great majority of Ga and Fanti resisted Islam and favoured the colonial school system, and many of them converted to Christianity. However, during the 1920s and 1930s the Ahmadiyya Muslim Mission spread throughout Fanti and Ga areas and, through its educational system, helped to convert hundreds to Islam. The Ga Aborigines Muslim Mission, founded in 1936, was one of many local missionary groups which began to change the religious situation of the southern Gold Coast. By independence in 1957, Islam had begun to make an impact in the Fanti and Ga communities. Also by this time a distinct split had developed between the 'immigrant' Muslims (often known as the 'Muslim Community') and the Ghana Muslim Mission which was led by Ga Muslims, many of whom had been raised in Christian households. In 1968 the Ghana Muslim Mission produced an offspring, the Supreme Council for Islamic Affairs, and in 1973 these organizations joined with the Muslim Community to form the Ghana Muslims' Representative Council (GMRC). However, the 'immigrant' and 'indigenous' Muslims spend more time arguing among themselves than co-operating on Islamic projects.[22] The absence of a unified Islamic movement and the failure to coordinate programmes with each other and with government agencies poses a serious problem for national development in the three nation-states.

Another issue of national concern in all three countries has been the preparation for and oversight of the annual pilgrimage to Mecca (*hajj*).[23] The pilgrimage is a fundamental religious obligation and a sign of high status for Muslims, and organizing the annual rite is a major undertaking. Furthermore, there are important economic, social and political implications for individual Muslims and NGOs. Substantial amounts of foreign currency must be obtained for transportation, accommodation and the purchase of goods for importation. The social contacts and political relationships developed from the pilgrimage greatly enhance the prestige and power of individuals and NGOs, which allow them to promote their interests and develop their programmes subsequent to this essential religious obligation. Often there have been allegations of fraud, favouritism and mismanagement in the arrangements for pilgrims by NGOs or by government agents. In 1988 some pilgrims from The Gambia complained that they had been 'abandoned' by their guides who seemed to be more interested in their own welfare.[24]

In 1987 Vice President A. B. Kamara charged that the Supreme Islamic Council had 'lost' $50,000 granted by the Sierra Leone government to be allocated to pilgrims.[25] The Federation of Sierra Leone Muslim Organisations was criticized by many Muslim clerics and missionaries as a creature of politicians, created only to gain control over external aid and the organizing of the pilgrimage. The principal organizer of the

Federation was Vice President A. B. Kamara, and the President of the Federation was al-Hajj Haroun Buhari who also was the Chief Press Officer of the government of President J. S. Momoh. In early 1990 the Federation had no secretariat, no budget, and was poorly prepared to plan for the *hajj*. Several leaders of Islamic organizations refused to co-operate with the Federation, because they viewed it as an effort by government politicians to interfere in their relations with Islamic states and agencies.[26]

For many years in Ghana there were allegations of embezzlement, misuse of air tickets and smuggling in connection with the *hajj*. The administration of President Hilla Limann (1979–81) appointed al-Hajj Imoru Egala and two members of parliament to evaluate several proposals for reorganization of the pilgrimage committee.[27] In 2007 the government of Saudi Arabia issued a strong warning about the ineptness of the Ghanaian *hajj* organization when hundreds of pilgrims arrived late for the rituals and more than 1000 were stranded after the pilgrimage ended.[28] One of the major difficulties in arranging for Muslims to make the *hajj* is that quite a lot of foreign exchange must be allocated for expenses for the journey, and often funds are misused or disappear altogether. Furthermore, pilgrims want to bring back as many souvenirs and gifts as they can and some may be sold in the markets. There is much room for corruption and favouritism, and several West African administrations have been heavily involved in the *hajj* arrangements.[29]

National and international relations of Islamic organizations

In the process of building schools and recruiting teachers, local and national Islamic associations become directly involved in the political arena. Many associations have been created to develop educational programmes that compete with or are parallel to the state system. Some of these, like Jamiyat Khuddam al-Islam in The Gambia or Basharia in Sierra Leone, are local self-help groups who receive little or no aid from abroad; while others, such as the Islamic Call Society with offices in several West African countries and the Islamic Research and Reformation Centre in Ghana, are closely connected with foreign agencies that provide considerable assistance.

Jamiyat Khuddam al-Islam has 13 schools and more than 1400 pupils in a programme on the north bank of the Gambia River; it builds its own schools and recruits and pays its own teachers; some of its textbooks are written by the director, al-Hajj Bashiru Darboe. The Islamic Call Society is based in Tripoli, Libya, and has several branches in West

Africa, including The Gambia, Sierra Leone and Ghana. The Islamic Call Society funds schools and teachers, supports agricultural and health projects, and provides scholarships in many subjects for study abroad. The Islamic Research and Reformation Centre has offices in Ghana (Nima, Kumasi, Tamale) and is supported by the 'Wahhabi movement' of Saudi Arabia. Some members left the Islamic Research and Reformation Centre in 1986, but it still operates a large school in Nima and has associates in Kumasi and Tamale. The 'Wahhabi movement' also supports organizations in The Gambia and Sierra Leone, but the government of Saudi Arabia assists many Muslim programmes in West Africa which are not affiliated with Wahhabiyya. For example, Saudi Arabia provided funds for the new central mosque in Banjul and for a college run by the Muslim Brotherhood in Magburaka, Sierra Leone.[30]

Muslim organizations are in constant contact with foreign agencies through correspondence, receptions for visiting delegations and missions sent abroad. These organizations actively seek scholarships for their most promising pupils and recruit foreign teachers to staff their school systems. Some, for example the Islamic Solidarity Association of West Africa in Serrekunda (The Gambia), the International Institute for Islamic Studies in Freetown and the Ghana Muslims Representative Council in Accra, operate what are essentially foreign affairs ministries. Organizations maintain extensive files of correspondence with foreign governments, with international agencies and with their own national governments. They frequently invite and host foreign delegations and regularly send representatives abroad to seek support. Their activities may become so well known that often their own governments will turn to them to look after a foreign delegation, or their representatives will be asked to accompany a government delegation sent abroad to seek foreign aid. The GMRC performed these functions in the 1980s. The GMRC position on this issue was stated clearly in a letter to the Young Muslims Association of Labadi: 'We may remind you that the Head of the Council is equal to any head of a Chancery, i.e. a diplomat, representing the Islamic State of Ghana.'[31]

The deep involvement of Muslim associations with foreign governments and agencies, and their interest in the welfare of Muslims the world over, leads them to take an active role in international affairs. The Gambia and Sierra Leone are both members of the Organisation of the Islamic Conference (OIC), and Muslim organizations advocate non-recognition of Israel and support for the Palestinian Authority and Palestinian land rights. The governments of Sierra Leone and The Gambia have supported OIC resolutions condemning Israel. Muslim leaders in

The Gambia said they were satisfied with the government's position. Gambian Islamic organizations do express solidarity with Palestinians when foreign delegations visit the country.[32] Recently President Alhaji Yahya Jammeh returned from the OIC meeting in Dakar and issued a strong denunciation of Israel's policies in Palestine and the lack of unified support from the Ummah.[33]

The Sierra Leone government of President Siaka Stevens firmly supported OIC resolutions and had close ties with Muslim Lebanese, but when President Joseph Momoh formed a new government in 1985 rumours began that there was a move to accord Israel diplomatic recognition. Several Muslim organizations reacted with letters to the Foreign Ministry, and influential leaders contacted President Momoh to express opposition. Dr Idriss Alami, Director of the African Muslims Agency, spoke with President Momoh and accompanied him on a state visit to Kuwait. He reported that President Momoh had stated that Israel would not have an embassy in Freetown during his presidency. Dr Alami also expressed strong support for Dr A. K. Koroma, the Foreign Minister, who firmly opposed an Israeli embassy.[34] Muslim officials argue that Islam incorporates political and economic issues, and that it is the responsibility of Muslims to express ideas and opinions about national and international affairs.[35] Sierra Leone also has a Muslim-Christian society dedicated to mobilizing support for the Palestine Liberation Organisation. The Sierra Leone-Palestine Friendship Society was founded in July 1987 to disseminate information about the treatment of Palestinians by Israel, to promote Palestinian statehood and to influence the government of Sierra Leone. It presented film, video and photographic shows at schools and at the Iranian Cultural Centre, and it sent pro-Palestinian resolutions to the government. In its publication, *SALPAL Magazine*, it emphasized that there are Christian as well as Muslim Palestinians, and it argued that the PLO is a political movement and not a terrorist organization.[36]

The situation in Ghana is rather different. It has a Muslim minority, and the Christian church membership comprizes more than half of the population. This seems to make members of the Islamic community even more active in supporting Palestinian independence. The Ghana Muslims' Representative Council has taken a strong stand on the Israel/Palestine question. In 1980 the GMRC organized a meeting to oppose the annexation of Jerusalem by Israel and to oppose the restoration of diplomatic relations. The GMRC also pointed out that Ghana received tens of millions of dollars in aid from the Saudi Arabian Fund, the Kuwaiti Fund, OPEC, the Arab and African Development Bank, and other Muslim-oriented agencies. Furthermore, Ghana relied heavily on petroleum imports from Algeria, Libya and other Arab countries. The GMRC advocated Israeli recognition of the PLO, repatriation of Palestinian people and restoration of lands taken during war.[37]

Internal Islamic agencies

The increasing frequency of contacts between Muslim organizations in West Africa and international Islamic agencies, and the development of educational and missionary programmes, have led to the establishment of offices, cultural centres and schools directed by representatives of several aid institutions. The Ahmadiyya Muslim Mission has been in West Africa for more than 80 years, and Egyptian cultural centres were set up in the 1950s and 1960s. However, the great majority of aid programmes and offices have been begun since the middle of the 1970s. This development is associated with the steep rise in petroleum prices and the greater control over production and marketing held by OPEC countries, but also it is a result of the resurgence of political Islam and the competition produced by this resurgence. A clear statement by the Ayatollah Khomeini on the political goals of Islam was published in *Echo of Islam*:

> Of course it is clear to every Muslim that worship within the realm of Islam has a vast meaning, and it is not confined to ritual prayer and fasting. Rather it encompasses solving problems of Muslims and removing the hegemony of the infidels from the Islamic states [...] Therefore, we also believe that the mosque is a place of worship in the broadest sense of the word.
>
> (August–September 1988, p. 9)

The nation-states most commonly found to be involved in funding and assisting in the operation of Islamic programmes in West Africa are Saudi Arabia, Libya, Kuwait, Egypt, Iran and, more recently, Pakistan and Sudan. They have embassies, consulates and cultural centres in various West African countries and all provide direct assistance, but these nation-states and the agencies which are based in them also work through local representatives. For example, the influence of Saudi Arabia is enhanced by the presence of an agent of the Muslim World League in Sierra Leone and by missionaries in Sierra Leone, The Gambia and Ghana who identify with the Wahhabi orientation.[38] The World Islamic Call Society, with its headquarters in Tripoli, Libya, is active in all three countries, where it has local offices and supports many organizations with funds, construction materials and scholarships. The Society also hosts international conferences to which it invites representatives from Islamic organizations.[39] The African Muslims Agency, which originated in Kuwait and has its principal office in South Africa, has sponsored educational and economic projects in all three nation-states.[40] Through cultural centres and the al-Azhar Mission, Egypt still plays an important role in Sierra Leone and Ghana. All of these of these international NGOs have been operating in the three states for more than 30 years. A more recent addition to the Islamic environment is the Tablighi Jama`at from Pakistan. While it has had little impact in Ghana and Sierra Leone, it has gained followers in The Gambia since missionaries arrived in the early 1990s.[41] Another recent NGO, which has its sub-regional headquarters in Banjul, is Munazzamat al-Da`wa al-Islamia from Sudan. It helps develop water supplies, health facilities, schools and mosques.[42]

Except for Lebanese immigrants, the Shi'a branch of Islam had little influence in West Africa. With the resurgence of political Islam in the 1970s this began to change. In 1975 a Shi'a missionary from Lebanon opened the Muslim Cultural Society in Freetown and began to recruit Sierra Leoneans to study in Iran. In 1977 Shaikh Houssein Chahade, a Lebanese educated in Roman Catholic schools in West Beirut and in the Islamic University at Qum, came to Sierra Leone to direct the Muslim Cultural Society. In 1981 a delegation from Iran, which included a Sierra Leonean student, came to discuss opening an embassy in Freetown. The Iranian embassy was established there in the same year. Since then several secondary and primary schools have been founded under the auspices of either the embassy or the Muslim Cultural Society, a few mosques have been built, a bookstore and library were established, and by 1987 the International Institute for Islamic Studies was founded to help coordinate the educational and cultural work of the Shi'a movement. The Institute operates the Islamic College (formerly the Ayatollah Hussein Montazeri College), a boarding school for 60 advanced students from West Africa in a complex specially constructed for the purpose. Some graduates are sent to Iran for further studies, if they want to go, and have the necessary qualification for further studies.[43]

The International Institute for Islamic Studies takes a very active part in Muslim affairs in Sierra Leone. In addition to participating in the formation of a new 'national' organization – the Federation of Sierra Leone Muslim Organisations – and strongly supporting Sunni clerics and students, it formed the Organisation of Islamic Unity in Africa (Sierra Leone) which sponsors Muslims to travel to Iran and which in

1983 organized a conference on *hajj* and Islamic unity in Freetown. The resolutions of the conference stated in part:

> Some Muslims live in disunity and their way of life is not in confirmation with the divine laws of Allah. The muslim sacred lands such as Palestine, Jerusalem, Lebanon, Afghanistan, etc. are under the occupation of invaders. The combatant muslims, ulamaas and mujahideen are under oppressions, terror, torture, mass killing and deportations [. . .]

> We consider the invasions of Afghanistan by the U.S.S.R., Iran by Saddam who is supported by the U.S.A., France and others; Lebanon by Israel and the mass killing of muslim minorities in different countries. These acts are considered to be plans to destroy Islam. In the same way we consider the South African regime as an instrument of world arrogance to suppress muslims in Africa. We therefore call all muslims in Hajj to discuss these problems and to take firm and clear positions according to the Holy Quran and the Sunnah. They should demonstrate their disapproval to these enemies of Allah, His chosen prophet (S. A. W.) and other muslim brothers.[44]

Despite the civil war in Sierra Leone (1991–2002), Iran continued its activities in Sierra Leone and seminary students were sent to the Imam al-Sadiq University in Qum. The programme in Sierra Leone provides funds for students to study abroad and also organizes celebrations for Muhammad's birthday, for Ali's accession to leadership (as leader of the *ummah* in 656 CE) and for the liberation of Palestine ('international day of the oppressed').

Although less active than in Sierra Leone, Shi'a missionaries are playing a role in the Gambia and attracting some followers:

> Ahl Albeit Islamic Association in collaboration with the Lebanese community in The Gambia, on Saturday, January 19, 2008, commemorated the martyrdom of Imam Hussein [. . .] Hundreds of Lebanese and other Muslims who gathered to acknowledge the good deeds of Imam Hussein in the restoration of Islam in a dangerous era, could not hold back their tears when Grand Imam Rabih Farhat recited numerous verses from the Holy Qur'an. Prayers were also offered to safeguard the Islamic ummah, the government of The Gambia, institutions as well as scholars.[45]

Iran and Gambia have solid governmental relations. President Yahya Jammeh personally invited President Mahmoud Ahmadinejad to attend the 7th summit of the African Union in Bajul in 2006. President Ahmadinejad spoke highly of the relationship and suggested that cultural and economic ties would be strengthened.[46]

In Ghana there have been many events organized by the Embassy of Iran and Ghanaian Muslim notables. In 1988 and 1989 there were international seminars on Islamic Unity Week at the Kumasi central mosque. The focus of these seminars was to celebrate the birthday of the Prophet Muhammad. Featured speakers came from the Islamic Republic of Iran, from the Tijaniyya in Senegal, and from Nigeria, Mali, Cote d'Ivoire, Sierra Leone and Ghana. The featured guest speaker in 1988 was Dr Muhammad ibn Chambas, the Provisional National Defence Council PNDC Deputy Secretary for Foreign Affairs.[47] As in Sierra Leone, Iran continued to devote considerable attention to Muslim affairs in Ghana. The Imam Hussain and al-Kawthar Foundations were established and several mosques were constructed. There is a large library collection, newspapers and a magazine in English, and many Islamic celebrations are organized. In 1988 the Madrasa Ahl ul-Bayt was opened to educate Ghanaians in Islamic studies. It was a residential school for 40 advanced students and planned to increase the intake to 100 scholars by 1991.[48] In 1988 the Ahl ul-Bayt Foundation began to develop the Islamic University College of Ghana which has since become the Islamic University of Ghana, located in East Legon, and affiliated with the University of Ghana in Legon and with the Islamic College for Advanced Studies in London. The Islamic University grants degrees in Business Administration and Religious Studies. It had its seventh graduation ceremony in 2007. Many of its graduates continue their studies in Syria and Iran. Its educational purpose is to train 'professional men and women who will [...] meet the highest standards [...] but will also be imbued with the commitment to serve in deprived areas in general and Muslim communities in particular'. In addition it seeks to produce 'mature individuals who have broad-base knowledge and appreciation of all existing religions for the purpose of encouraging understanding and dialogue between different religions and cultures'.[49] Another organization that promotes Iran's cultural and religious ties in Ghana is the Islamic Culture and Communication Organisation. Among its principal objectives are the 'Promotion and consolidation of cultural ties with various countries and Islamic nations in particular aimed at cultural exchange and presenting the true nature of the culture and civilization of the Islamic Iran'; to gather social, political and legal information about the 'status of Shiites throughout the world'; to protect the rights of Shi'a followers; to organize 'gatherings, festivals, propagation and cultural exhibitions and public ceremonies abroad'; and to develop Iranian studies and Persian language programmes in foreign universities.[50]

Islamic NGOs and economic development

This chapter has indicated several ways through which Islamic associations have contributed to the economy of the three nation-states studied, especially in the areas of education and the provision of social welfare (health clinics and immunization programmes, orphanages, the distribution of food and clothing, conferences about female genital mutilation and family planning, reference libraries, burial services, agricultural programmes, construction, transport and administration). The economic value of these activities have not, to my knowledge, been calculated but they provide services, education and employment for large numbers of Muslims and non-Muslims in the three nation-states. Many of the *jama`at* and larger associations attempt to organize economic schemes to improve the welfare of their members. The Islamic Solidarity Association of the Gambia built a medical clinic in Serrekunda, while the Al Atharee Association provided a reference library for Kartong.[51] In Ghana many associations are active in development schemes. One in particular, the Islamic Council for Development and Humanitarian Services, is a successful local NGO that promotes social welfare projects throughout the country.[52]

The situation is the same in Sierra Leone where one finds many local and national agencies that provide a variety of services. The Muslim Brotherhood (a national organization unaffiliated with the Egyptian Brotherhood) and the Ansarul Islamic Mission (a branch of the Tijaniyya *tariqa*) operate schools and provide social services such as counselling and legal advice. One local movement, similar to the Ahl al-Sunna wal-Jama`at in Ghana, was founded by Shaikh Bashar Sankoh Yilla in the 1970s when he left the large Temne central mosque headed by Shaikh Jibril Sesay. Shaikh Bashar, educated in Saudi Arabia, advocated an austere form of Islam and opposed the use of drumming and Islamic amulets and potions. He felt that ostentatious living was un-Islamic and community funds must be used to alleviate poverty and for the development of the community. He lived very simply and treated his followers like an extended family. With the support of powerful patrons within the Temne community he established a religious centre in Freetown, attracted many young Muslims to his movement and collected substantial *sadaqah* (voluntary alms) to be used for housing, schools and social services.[53] The Basharia grew in popularity, spread from Freetown to the interior and continued to thrive after his death. The current head of Basharia is *al-hajj* Abu Bakarr Kamara who supports the programme of Shaikh Bashar and is an advocate of honest and transparent governance. He is a member of the National Accountability Group-Sierra Leone whose

> Vision is to have a country in which local and national government are accountable to the people, budget allocation, resources, management and public expenditure are open and transparent, and business and the daily lives of the people are free from corruption and poverty.[54]

Historically, Muslim communities have relied upon voluntary giving (*sadaqah*) and prescribed alms (*zakat*) to finance social welfare programmes. In addition, the religious endowment (*waqf*) was created to support larger-scale projects. The funds were used to build Friday Mosques (*masjid jaami`*), hospitals, law schools, universities, accommodation for travellers, to purchase lands for farms and religious centres and to create foundations to oversee many of these projects. The poorer members of the communities benefited from medical care, education, and subsidies to make life easier. Widows, orphans and the elderly poor were especially identified as worthy of assistance. Sufi *turuq* often emphasize the love (*hubb*) that Allah feels when a Muslim pursues a life of service to others.[55] These concepts are well known in the three nation-states; *sadaqah* and *zakat* are collected and used to benefit the communities and the establishment of *awqaf* (Islamic endowments) has been contemplated. Also *al-shari`a* recognizes a number of instruments for economic development and finance, including the historic use of the *bayt al-maal* (treasury or central bank); and Islamic banks and other financial institutions have begun to play a role in development schemes.[56] Some economists and financial advisors have recognized the value of Islamic banking under the present economic conditions.[57] As we have seen, some local and national NGOS have been successful in obtaining their own resources to support their projects, but the economic conditions in the three states are insufficient for self-generating and self-sustaining development. Furthermore, without a unified *ummah* necessary for coordination of the collection and distribution of Islamic funds in these states not enough money can be generated for large-scale projects. With few exceptions, domestic Islamic organizations seek aid either directly or through their governments from external NGOs, foreign governments or international agencies.

Since the 1970s there has been a remarkable increase in the number and activities of Islamic NGOs in Africa. For the three nation-states examined here most of the significant NGOs have been discussed. In addition, states not associated with the colonization of Africa have become more and more active in their political, cultural and economic affairs. Except for Egypt, which took an interest early on through the work of Al-Azhar University, the activities of external governments have been fuelled by substantial income from the petroleum industry and were further stimulated by the Islamic revolution in Iran which led to competitive efforts principally by Saudi Arabia, Libya and Iran.[58] Their governments have established organizations based on endowments which allow them to participate directly in the affairs of the *jama`at* of West Africa, and they have also developed close economic ties with The Gambia, Sierra Leone and Ghana during the past 30 years.

Governments and NGOs also work within international agencies to acquire and distribute funds. A primary emphasis has been placed on the role of education in creating a more prosperous and fulfilling future for these three states. While other types of development programmes are being pursued (health, housing, agriculture), Muslim leaders recognize that to provide children – especially girls – with productive skills will go a long way towards alleviating poverty. Consequently the governments and Islamic organizations in all three countries have sought funding from international agencies to develop or – in the case of Sierra Leone – to rebuild their educational facilities. The Organization of the Islamic Conference, the Islamic Educational, Scientific and Cultural Organization and the World Bank have been especially supportive in providing funds for infrastructure, teacher training, text books and supplies.[59] At its international meeting in March 2008 held in Dakar, Senegal, the Organization of the Islamic Community, along with more than 60 Islamic NGOs, proposed to launch programmes 'to reduce inequality amongst Islamic states' and to 'play a greater role in providing humanitarian assistance to Islamic countries'. Foreign Affairs Minister Cheikh Tidiane Gadio of Senegal said, 'The plan is not just to provide "zakat" [charity] to poor states but a genuine mechanism by which the wealth of Islamic states can be more equal.' An anti-poverty fund of US$10 billion was proposed in May 2007. The combined OIC–Islamic NGO conference

closed with a joint statement calling on governments throughout the Islamic world to support humanitarian NGOs in their countries. The OIC pledged to create a centre to analyse humanitarian needs in OIC countries. It also said it would establish more formal links with NGOs.

Although Ghana is not a member of the OIC, Islamic NGOs from that state are welcome to apply for economic assistance from the OIC and from the many other international Islamic agencies that are funded by governments and private organizations.[60]

Conclusion

The great increase in and intensification of international Islamic contacts and internal activities caused governments in The Gambia, Sierra Leone and Ghana to try various means to bring them under control. Islamic organizations frequently do function as independent agencies. They operate schools, clinics, libraries and institutes, they organize educational and missionary seminars, they sponsor pilgrimages and visits to conferences abroad, and they seek financial and missionary assistance from foreign agencies and even governments.

Governments in these West African countries have sought to register all organizations, to establish national Islamic federations, and to place the activities and budgets of all organizations under government control. Such attempts have failed for several reasons:

- Governments in the three countries do not have economic or clerical resources necessary to operate federations.
- The Muslim communities are too disparate for truly inclusive, national federations to be formed.
- Islamic organizations are suspicious of the motives of politicians whom they accuse of power-grabbing and avarice.
- Competition between Islamic organizations for members and resources is intense.
- International agencies and foreign governments reinforce this competition by seeking allies amongst the Muslim organizations in pursuit of their own agendas.

West African governments have never succeeded in convincing Islamic organizations to subordinate their individual identities to a national federation. Although often they have been frustrated by lack of recognition or by the slow pace of aid projects, Muslim organizations operate aggressively and openly to serve their followers and to seek assistance from Islamic governments and agencies. As examination of Muslims and their organizations have shown, they have successfully built and expanded Islamic space in the three West African states and have also launched political, economic and social initiatives to increase their influence and leverage in national affairs. The many self-help projects undertaken by these organizations have helped to provide economic opportunities in the three countries but the absence of unity and the failure to coordinate their activities with government programmes have inhibited the development of a national economic plan.

The intensity of Muslim missionary and educational activities seems to be greater in Ghana than in either Sierra Leone or The Gambia. This may be because Muslims in Ghana, who have been distinctly in the minority during the colonial and postcolonial era, perceive the need to aggressively pursue their objectives and to develop a well-defined Islamic space for their communities. Ghana is by far the most populous of the three countries and has a tradition of strong political organization. Not only is Ghana highly politicized, but its leaders have been especially active in international affairs for several decades. Muslims have played prominent roles in all Ghanaian administrations from Kwame Nkrumah to the present and have sought to increase Muslim leverage in political, economic and social spheres.[61] Muslims developed their institutions at an earlier period in The Gambia and Sierra Leone, and during the colonial era Islam was the predominant religion, although Christian Africans played important roles, especially in Sierra Leone. Muslim leaders successfully expanded Islamic space and asserted themselves politically. In the postcolonial era they have increasingly taken significant leadership positions in political, economic and social affairs. Although Islamic organizations are competitive in Sierra Leone and The Gambia, they are working to promote their interests within a solidly Islamic environment. As members of the Organization of the Islamic Conference they are recognized as Muslim states with an interest in promoting Islamic ideas, values and programmes, but external Islamic agencies and foreign governments also view Ghana, with its large population and significant international reputation, as important base for the expansion and strengthening of their interests in West Africa, and greater resources have been provided to Islamic organizations there. After long periods of political instability and economic decline both Sierra Leone and Ghana have returned to systems of regular and competitive elections. In Ghana after eight years of governance by the New Patriotic Party under President John Kofi Agyekum Kufuor and Vice President Alhajj Aliu Mahama, the National Democratic Congress won a narrow victory and in January 2009 John Atta Mills and John Dramani Mahama formed a new government. In September 2007 President Tejan Kabbah of the Sierra Leone Peoples Party relinquished the administration of government to Ernest Bai Koroma, the candidate of the All Peoples Congress. The Muslim communities in both countries are well represented in Ministerial positions.[62]

Islamic missionary organizations are not behaving in an unusual manner. Roman Catholic, Anglican, Wesleyan, Baptist and other missionary movements have been active in Africa for centuries. Many of their activities also emanated from educational and political programmes. Muslims – foreign and domestic – are pursuing goals similar to other religious bodies (including the Muslim merchants and missionaries of 13 centuries ago in Africa) which have operated in Africa. However, goals pursued by competing organizations, agencies and governments can lead to destabilizing consequences. Several Muslim African officials expressed fear that external disputes might affect the ability of Muslim organizations to work within the internal political system. They felt that Islam might become less effective as a model for economic development and social change but they saw no easy solution to this problem, as Muslim organizations require external assistance in order to pursue their goals. It is likely that Islamic and national conflicts and divisions will be projected onto the politics of Islam in West Africa.

The invasion of Kuwait by Iraq and the potential conflict between Iraq and Saudi Arabia adversely affected Islamic educational and development projects in West Africa. Although Iraq has not been a significant source of funds, Kuwait and Saudi Arabia provide substantial aid. Even without a major war in the Middle East – with subsequent damage to oil production and income – the resources of Kuwait and Saudi Arabia were diverted to the immediate problem of dealing with Iraq's military actions or internal instability. Throughout the 1990s Libya, Iran, Egypt, Saudi Arabia and Kuwait continued to provide support for *da`wa* in West Africa, but economic decline imposed restraints compared to the 1980s. However, the sharp increase in oil revenue and intense rivalry between Saudi Arabia and Iran has tended to stimulate competition for the 'hearts and minds' of Muslims. The continuing conflicts in the Middle East and the 'war against terrorism' which has been launched in Africa through US military initiatives add to uncertainty in the region. A wider war in the Middle East, an attack on Iran or major military interventions in Africa would likely lead to catastrophic economic consequences.[63]

Notes

1. For an analysis of this development see Skinner (2009).
2. While the populations of each country were influenced by missionaries from Mande-speaking groups, Hausa missionaries were quite significant in the Gold Coast/Ghana and Fula teachers were very important in Sierra Leone. The Gambia is the smallest of the three nation-states, with a population of almost 1.7 million of which about 90 per cent are Muslims. Sierra Leone has a

population of more than 5 million, and estimates of the Muslim population are around 65 per cent. Both The Gambia and Sierra Leone have small Christian populations. Ghana is by far the largest of the three nation-states with a population of more than 20 million. The most recent census placed the Muslim population at around 16 per cent but other estimates range between 20 and 45 per cent, the latter surprizingly coming from the Catholic Secretariat in Accra. Soon after the First World War, the British colonial administration judged that Muslims made up at least 15 per cent of the population. With substantial conversion to Islam amongst the Ga, Fanti, Ashanti and other southern Ghanaians the Muslim population in 2008 could possibly be as high as 30 per cent (Esposito, 2003, p. 93).

3. By the first decade of the twentieth century, 'modern' Muslim schools had been established in all three territories. For additional discussion of the period before 1950, see (Skinner, 1976, 1983, 2009).

4. In April 1983 the Gambia Muslim Association, with the assistance of the World Assembly of Muslim Youth and the Organisation of the Islamic Conference, held a nine-day international youth camp. The programme included presentations on law, education, the spread of Islam in West Africa, self-reliance, advocacy of Islam, the role of Muslim youth in national development, and others. Between 1963 and 2007 the Gambia Islamic Union built 136 schools: Sheriff Barry, 'Foundation Stone for Islamic School Laid', *The Daily Observer* (Banjul), 18 June 2007, www.allafrica.com. Ideas of self-reliance and national development could be developed more in the general text.

5. Interview with al-Hajj Abdullai Jobe, Banjul, 18 November 1988. Al-Hajj Jobe is a university-educated specialist in Islamic law and philosophy and has written two treatises in Arabic.

6. In 1982 the village of Mabrouk, near Accra, had a beautiful new school and mosque.

7. Interviews: Ghana; for a study of Nima see Verlet (2005).

8. Report by al-Hajj Rahimu Gbadamosi, Northern Region Director of Education, April 1982.

9. Later in the chapter we discover that organizing the pilgrimage has been a basis for conflict among Muslims, especially in Ghana and Sierra Leone. The pilgrimage has been an important source of revenue enhancement, corruption and power politics.

10. Report by Hassan B. Bangura, 15 October 1988; for a study of the Headman administrative system see Harrell-Bond, Howell and Skinner (1978).

11. While there is no accurate count of the number of Islamic organizations, in each country there are hundreds, if not thousands, of associations of various types, ranging from individual mosques or schools to national federations.

12. The Organisation of the Islamic Conference has its headquarters in Jiddah, Saudi Arabia. It functions to fund and coordinate economic programmes and to promote unity on issues which are considered important to Muslims around the world; see www.oic-oci.org.

13. When the Muslim Action Party was founded in the Gold Coast (Ghana) during the 1950s Kwame Nkrumah had al-Hajj Imoru Egala, a northern Muslim politician, form a rival youth organization, and MAP was absorbed into the

Convention Peoples' Party. The Gambian Muslim Congress also was eliminated by competition with the more national Progressive Peoples' Party. In the three countries Muslim leaders have opted to participate within the secular political party system and to form alliances with non-Muslim leaders in order to gain political leverage.

14. The Islamic Call Society also funds the World Council for Islamic Da`wah, which has an executive council of 32 members from all areas of the world: Interviews: Drammeh; and www.islamic-call.org. It continues to fund and support Islamic projects in Africa and elsewhere and sponsors international conferences: 'Mercy for Mankind'.

15. Interviews: Jallo, Banjul. Similar and very serious divisions over organization control and funding have occurred in Sierra Leone: Interviews: Fofanna; Abdul Qadri Koroma; and Drammeh.

16. The Supreme Islamic Council itself had been created because of a political party conflict. When the Sierra Leone Peoples' Party (SLPP) controlled the government, the Sierra Leone Muslim Congress, led by a leading SLPP politician, M. S. Mustapha, was a powerful organization. The SLPP was overthrown in 1967 and replaced by the APC, and the Supreme Islamic Council, led by S. A. T. Koroma, emerged as an important organization. However, 20 years later there are many rival organizations, each with its own base of support: Interviews: Fofanna; U. N. S. Jah; and Y. A. T. Bangura.

17. Dr A. K. Koroma began this project when he was Minister of Education. Estimates for the amount expended on the new college range between $1,000,000 and $2,500,000: interviews with al-Hajj M. M. Fofanna, Freetown, 2 June 1987; al-Hajj U. N. S. Jah, Freetown, 3 December 1988; and Bilal Kargbo, Freetown, 3 December 1988. A substantial sum – perhaps another $1,000,000 – would have been required to complete the construction and equip the college. Al-Hajj M. L. Sidique, the Secretary General of the Sierra Leone Muslim Brotherhood and Minister of Labor (and formerly the Minister of the Interior), planned to visit Kuwait, Saudi Arabia, Egypt and Libya to acquire the funds to complete the project: interview with M. L. Sidique, Freetown, 6 December 1988. The Islamic Call Society allocated $600,000 to the Sierra Leone Muslim Brotherhood for work on the college in 1989: *New Citizen*, Freetown, 26 August 1989. The civil war during the 1990s in Sierra Leone prevented the completion of this project.

18. GMRC File 8a, p. 5 (draft memorandum to President Hilla Limann); File 9/10, p. 3 (GMRC press release, 18 September 1980); File 11, p. 5 (letter from Ghana Muslim Community, 2 June 1981); File: Muslims' Consultative Committee (resolution passed by Muslim Preachers' Association regional meeting, Takoradi, 13 November 1981).

19. President Sir Dawda Jawara requested that a new mosque committee be selected in order to make it more representative, but this had not been accomplished as of December 1988: Interviews: Drammeh; Joof; Ndow; and Jobe. The mosque has been put into service, but the committee remains dominated by Wolof notables with some Aku representation: communication from Ibraima C. Jallo, Banjul, 1 May 1990. Historically in the Gambia, villages developed according to ethnic identity: Mandinka, Serakhule, Jola, Fula, Wolof, although there has been intermarriage between the communities.

20. Harrell-Bond (1978) shows that the label 'ethnic mosque' does not mean that all who pray there belong to one ethnic group; it means that the mosque is dominated by a particular ethnic group who raised the funds to have it built.

21. While some were unconcerned about it, others expressed hostility. They felt that as it had the appearance of a central or national mosque the imam should be Sunni. The Lebanese community in Sierra Leone is very well established and has played an important role in both economic and political affairs. Shaikh Chahade, who left Sierra Leone in 1987, formerly provided funds to many of the imams of Sunni mosques in Freetown; however, the remnant of his mission continued to function, and his brother operated the al-Hadi Bookshop and Library while Shaikh Ahmed Tijan-Sillah directed the Muslim Cultural Society. Interviews: Bilal Kargbo; Alami; Hassan Bangura; Tijan-Sillah; Fofanna; Mohammad Ramadan Bah; Chahade.

22. Interviews: Classpeter-Williams; Collison-Kofi; Darpoh; Okine; Otoo; Ahmed; Futa; and Dretke (1968). Ethnic and sectarian divisions also prevented the opening of the central mosque at Abossey Okai in 1982. Different organizations claimed authority over the new mosque, and there was no consensus about who would choose the imam and other officials to manage it.

23. To become an *al-haajji* (often written Alhaji) is an honour and commands great respect.

24. Under the administration of President Yahya Jammeh the government has attempted to control the organization of the *hajj*. President Jammeh also attempted to gain support for his regime by personally sponsoring large numbers of pilgrims: Pa Malick Faye, 'Pilgrims', *The Daily Observer*, 26 November 2007: www.observer.gm.

25. This charge and others made by Vice President Kamara in January 1987 were seen by leaders of the Supreme Islamic Council as part of the government's plan to undermine the Council and replace it with the Federation of Sierra Leone Muslim Organizations, which was formed under government patronage in May 1987: Report by Bangura, 15 August 1988.

26. Interviews: Fofanna; Hassan B. Bangura; Alami; Swarry; Kabia, Minister of Rural Development, Social Services and Youth; Sasso; U. N. S. Jah.

27. GMRC File 11: Letter from al-Hajj A. B. Futa and Malik Abdul Mumuni, 2 June 1981. Al-Hajj Egala was Kwame Nkrumah's agent to bring Muslims into line behind the CPP in 1956. Al-Hajj Egala was in jail after the coup against Nkrumah in 1966 and was barred from holding elected office, but he was a founder and patron of President Limann's PNP and considered the power behind the president. He died in 1981: *Daily Graphic*, 2 April 1981. The second Rawlings government (31 December 1981 to the present) suspended the *hajj* committee and refused to allocate any funds for 1982.

28. Daily Guide, 'Saudi Sanctions Ghana Over Hajj', 22 January 2008: www.modernghana.com/news; 'Hajj Committee Must Refund Monies', 16 April 2008: www.peacefmonline.com.

29. One Muslim association in Ghana complained that every government since independence had its own Muslim council to act as a mouthpiece, and the association argued that it was time for Muslims to co-operate in order to strengthen the *ummah* in Ghana and to prevent further political interference: GMRC File, Muslims' Consultative Committee,

letter from the Muslim Preachers' Association, Takoradi, 11 November 1981. This dispute continues in the twentieth century with many complaints about missing funds, poor administration and favouritism: www.modernghana.com. (news for 22 January 2008, 11 December 2007, 24 June 2007); www.gambianow.com/news. (24 December 2007); www. allafrica.com/stories/00608020746. html (Corrupt hajj agency owes pilgrims $94,000.).

30. This information is based on interviews and NGO files: Darboe; Drammeh (al-Hajj A.C. Banding Drammeh, president of the Supreme Islamic Council in the Gambia); Aisha; Fofanna; GMRC File 3A/1982; Darpoh; Hassan Zacharia; Husseini Zacharia; Yaya; Umar Ibrahim Imam (Sheikh Umar, the national imam for Ahlus-sunnah wal-jama'ah, the principal Wahhabi-influenced organization in Ghana) and Kotey.

31. GMRC File 11, 20 February 1981. Every constitution of a Muslim organization I have collected contains a provision that the organization will act as liaison with foreign governments and agencies. Furthermore, many officers of Islamic organizations have been important governmental officials or administrators and have pursued both Islamic and government policies.

32. One informant said that Jews and Muslims should be able to live together peacefully as they had done in the past. There is a PLO office in Banjul, and reportedly it obtained $70,000 for the Islamic Union. President Dawda Jawara was considered a moderate in international affairs, and one did not read hostile statements about the Israeli-Palestinian situation in the government press.

33. 'Be True, Unite!' and 'So True!' (Editorial), The Daily Observer, 17 March 2008: www.observer.gm. Since taking power in 1994 President Jammeh has become a principal spokesperson on Islamic affairs in the Gambia, although not without criticism from Gambians: 'Gambians wish Jammeh well on Eid-al-Fitr, but...', Gambia Post, 14 November 2005: www.gambiapost.net: 'Jammeh should know that all the money he is mismanaging in this country is tax payer's money.' President Jammeh, whose personal Islamic credentials were slight when he came to power, has tried to control Muslim affairs in his country through the Supreme Islamic Council. He also ordered the construction of a State House Mosque which has an Imam and Mosque Committee separate from other Muslim communities in The Gambia: Musa Ndow, 'State House Mosque Committee Donates', The Daily Observer, 5 February 2008, www.allafrica.com. Many Muslim scholars and elders have resisted centralized authority, and President Jammeh has had minimal success in dominating their activities: see Darboe (2007).

34. Speaking at the third United Nations African seminar on Palestine, Dr Koroma said that it would not be appropriate to have an Israeli embassy in Freetown before the rights of Palestinians are settled: West Africa, 14–20 May 1990, 821; communication from Hassan B. Bangura, 17 May 1990.

35. Interviews: Alami; U. N. S. Jah; Y. A. T. Bangura. Dr Alami, a Sierra Leonean of Moroccan descent, was involved in many of the Islamic organizations since 1979 and worked with the African Muslims' Agency (AMA) for several years. The AMA was founded by Dr Abdu Rahman Sumait in Kuwait. It promotes education, communications and other missionary work in several African countries.

36. Hassan B. Bangura, 'Research Report on the Activities of the Sierra Leone/Palestine Friendship Society', August 1990. The government of President Momoh has recognized the state of Palestine, and there is a Palestinian embassy in Freetown.

37. GMRC Files 8a and 11. Of course, economic aid is a fundamental issue for the poorer countries of the Third World, and with the rise in petroleum prices during the 1970s many Muslim states had an opportunity to increase their support for general economic projects and specific Muslim programmes. The Gambia, Ghana and Sierra Leone have received substantial assistance from many Arab/Muslim governments and international agencies such as the Islamic Development Bank, Islamic Conference Organisation, Islamic Solidarity Fund, the Muslim World League, the Higher World Council for Mosques, the World Association of Muslim Youth and OPEC. However, petroleum prices cut both ways, and many Third World governments have been dissatisfied with the amounts of aid provided, given the high cost of petroleum products. In addition to high energy costs, another factor that subtly undermines African-Arab relations is the perception by some African clerics that Arab Muslims are too arrogant or think that Islam is an Arab religion. A third factor that mitigates African-Arab solidarity is the lack of unity among Muslims in Africa. There are many rival groups, each of which hopes theirs will become the prominent organization in their region or nation-state.

38. The Ahlu-s-Sunnah wa-l-Jama`at in Ghana and the Basharia Mission in Sierra Leone are two of the prominent Wahhabi-influenced movements.

39. Names of WICS Centres Around the World: www.islamonline.net/the-wics; 'Mercy for Mankind', 7th General Conference: www.islamonline.net/mercyforworlds; 'Training for Teachers Ends at Maummar Gathafi Mosque', *The Point Newspaper*, 30 August 2007: www.thepoint.gm; 'Rumours Around Libyan Gift', *The New Citizen*, 10 March 2008: www.christiantrede.com/webdesign/clients/newcitizen/localnews.

40. www.africanmuslimsagency.co.za; Abidou Rahman Sallah, 'Gambia: African Muslims Agency visits Islamic Schools', *Daily Observer*, 26 April 2007: www.muslimnews.co.uk/news/news.php?article=12654.

41. Janson (2005). While there have been individual missionaries in Ghana and Sierra Leone who have not formed communities, there is a very active community of Africans who have joined the movement in The Gambia.

42. Aboubacar Abdulah Senghore, 'Islamic Endowments in the Gambia', *The Point Newspaper*, 16 and 26 August 2005.

43. Interviews: Chahade; Ramadan Bah; Tijan-Sillah; Bilal Kargbo; Fofanna; Jibril Sesay; Hassan Bangura; Irawani; A. B. Kargbo; Alami. Shi'a missionaries have established a similar programme in Ghana, and there are missions in other West African countries. In Accra, the Madrasa Ahl al-Bait was a residential school for 40 advanced students in 1989 and intended to expand to more than 100 students. Interviews: Tabataba'i; Bansi; Husseini Zacharia.

44. The quotes are taken from a document produced by the Organisation of Islamic Unity which sponsored the conference in Freetown from 26 to 31 July 1983. Attending the conference were 25 delegates from Sierra Leone and 200 delegates from other Muslim countries. The delegates included both Shi'a and Sunni representatives. The International Institute

for Islamic Studies, the Embassy of Iran and the Cultural Consulate remain very active in Islamic affairs in Sierra Leone: Alpha Amadu Bah, 'Stake Holders Explain Prophet's Strategies in Peace Building', 17 January 2008: www.sierraexpressmedia.com; Abu Bakarr S. Tarawally, 'Eid-ul Adha Feast Celebrated', 22 December 2007: www.sierraexpressmedia.com.

45. Hatab Fadera, 'Ahl Albait Islamic commemorate martyrdom', 25 January 2008: www.wow.gm.

46. P.I.R.I. News Archive, 02 July 2006: www.president.ir.

47. *Peoples' Daily Graphic*, 28 October 1988 and 13 October 1989.

48. Interviews: Tababa'i; Bansi; Husseini Zacharia.

49. Islamic University-Ghana, 'Our Mission': www.islamicug.com. In 2006 Dr Gholamreza Rahmani Miandehi was appointed the new President of the Islamic University-Ghana. Ahl ul-Bayt claims there are one million Shi'a residing in Ghana.

50. www.iranculture.org

51. Interviews: Ibraima Jallo; also see Nfamara Jawneh, 'Islamic Reference Library opened in Kartong', 05 February 2008: www.wow.gm/news/topic/islam/rss.

52. The Council was founded in 1991 by a Ghanaian businessman, Shaikh Mustapha Ibrahim, who in 1980 established the Islamic Book Development and Translation Council to provide texts for Ghanaian Muslims (interview, Accra, 18 September 1980). It was a very successful endeavour. Shaikh Mustapha has been able to finance the Council's projects and to support other Islamic organizations through *zakat* collections within the Ghanaian community. For a very useful study of the Council and other Islamic associations see Weiss (2007), especially chapter V.

53. Interviews: Ramadan Bah; Hassan Bangura; U. N. S. Jah; Sankoh Yilla; Mohammad Kamara,. Shaikh Bashar Sankoh Yilla was a protégé of Shaikh Jibril Sesay, imam of the Temne Muslim community, who arranged to have him study in Saudi Arabia.

54. www.accountability.sl. Sheikh Abu Bakarr Kamara was educated in Sudan and Egypt. He is National Chief Imam of the Basharia Jamat and Emir of the United Council of Imams of Sierra Leone.

55. Sufi *turuq* (Orders) are active in all three nation-states: Qadariyya, Tijaniyya and Muridiyya in The Gambia; Tijaniyya in Sierra Leone; and Qadariyya and Tijaniyya (two branches) in Ghana. The Tijaniyya are especially active in economic and social affairs.

56. Aboubacar Abdulah Senghore, 'Islamic Endowments in The Gambia', *The Point Newspaper*, 16 August 2005: www.thepoint.gm/muslim_hands; Dr. Omar Jah Jr., 'Issues in Islamic Economics,' *The Point Newspaper*, six parts, April–June 2006: www.thepoint.gm/muslim_hands; Interviews: Mohammad Jallo, Sanu Barry, Abu Bakar Jallo, Sajalieu Bah; Deegan (1995); Linden (October 2004); Weiss (2007). A few of these financial arrangements are *al-mushaarakah* (equity participation), *al-mudaarabah* (limited partnership), *al-mufaawadah* (equal partnership) and *al-muraabahah* (bank loan contract). Some useful websites for reviewing Islamic banking and finance are: www.islamic-finance.net; www.cie.com.pk; www.islamic-world.net/economics; www.albalagh.net/Islamic_economics; www.islamiccenter.kaau.

edu.sa; www.soundvision.com/Info/money/islamiceconomicterms.asp; www.
islamic-foundation.org.uk.

57. Umesh Desai, 'Islamic finance offers safeguards to investors' in *International
    Herald Tribune*, 5 April 2008, p. 15. This has become increasingly clear during
    the international financial crisis in 2008 and 2009.

58. Mu'ammar Qadhdhafi, as a Muslim African leader, has taken a particularly
    keen interest in the affairs of West African states.

59. Dr Omar Jah Jr., 'The Importance of Early Child Islamic Education', *The
    Point Newspaper*, 7 October 2005: www.thepoint.gm/muslim_hands; 'Mus-
    lims Urged to Make Education Their Topmost Priority', *Public Agenda*,
    4 October 2004: www.ghanaweb.com; Holgar Weiss, *Begging and Almsgiv-
    ing in Ghana*; The Organization of the Islamic Conference, the Islamic
    Educational, Scientific and Cultural Organization, and the United Nations
    Children's Fund, *Investing in the Children of the Islamic World*, UNICEF,
    November 2005; Document of The World Bank, 'Project Appraisal Docu-
    ment on a Proposed Grant in the Amount of SDR 15.1 Million (US$ 20.0
    Million Equivalent) to the Republic of Sierra Leone for a Rehabilitation of
    Basic Education', 28 January 2003 (Report No. 23413-SL). Contained therein
    is a contract for the Provision of Basic Education in the Tonkolili District,
    and the name of the contractor is the Sierra Leone Muslim Brotherhood.
    After the decade-long upheaval in Sierra Leone there were few functional
    schools remaining in the country. For recent educational initiatives in The
    Gambia see The Gambia National Commission for UNESCO, Education –
    ISESCO: www.gambia.comnat.unesco.org.

60. IRIN Africa, 13 March 2008: www.irinnews.org. For a complete report on
    the OIC conference see www.oic-oci.org. The 'sovereign wealth funds' and
    the treasuries of petroleum-exporting states were awash in 2008 with wealth
    available for distribution or investment, but it is unclear whether it will
    be spent wisely, who will benefit from the funds or how much funding
    will be available in the future: Kathryn Hopkins, 'Sovereign wealth funds
    "could be bigger than US economy by 2015"', *The Guardian*, 28 April 2008;
    'Gulf Economies: How to spend it', *The Economist*, 26 April–2 May 2008,
    pp. 37–8, 40. The Islamic Educational, Scientific and Cultural Organization
    works closely with the OIC, WICS, UNESCO and the World Bank to help
    organize and fund development programmes.

61. In the administration of President John Agyekum Kufuor, Vice President
    Alhaji Aliu Mahama played a leading role in the Muslim community.
    Other Muslims who have held ministerial positions include Alhaji Boniface
    Abubakar Siddique, Hajia Alima Mahama, Alhaji Mustapha Ali Idris and
    Sheikh I. C. Quaye. During the administrations of Jerry Rawlings there were
    several Muslim officials and personal advisors.

62. Since 1982 the long period of relative political stability in Ghana has
    enhanced its attractiveness for economic and missionary endeavours, and
    the end of civil war and economic turmoil in 2001 has given hope to Sierra
    Leoneans to rebuild their economy in a more peaceful environment. While
    both John Atta Mills and John Dramani Mahama are Christians (Mahama
    comes from a prominent Muslim family in northern Ghana but is a convert
    to Pentecostalism), they have been reaching out to the Muslim communities

in public ceremonies, and the Minister of Foreign Affairs (in April 2009) is Alhajj Mohammed Mumuni. In Sierra Leone the irony is that the Christian President Ernest Koroma has a more predominantly Muslim cabinet than his Muslim predecessor, Tejan Kabbah. Since 1994 the Gambian President Alhaji Yahya Jammeh has been pursuing his own Islamic agenda as 'head' of the Muslim community, but not without some rather well-hidden dissension.

63. The creation of AFRICOM (United States' Military Africa Command) and the focus on the 'war on terror' contribute to economic and political uncertainty in Africa (and elsewhere). Initiatives by the US government in 2008 and 2009 indicate a heightened military interest in Africa (including Ghana, where AFRICOM may be based). The initiatives of China and India in Africa also may dramatically affect political and economic landscapes in Africa. This chapter was written before the extremity of the global economic crisis had developed, and changes in funding by governments, NGOs and international agencies are unclear.

References

Darboe, M. N. (2007), 'The Gambia: Islam and Politics', in W. F. S. Miles (ed.), *Political Islam in West Africa: State-Society Relations Transformed*, Boulder: Lynne Rienner Publishers, pp. 148–58.

Deegan, H. (1995), 'Contemporary Islamic Influences in Sub-Saharan Africa: An Alternative Development Agenda' in E. Watkins (ed.), *The Middle East Environment*, Cambridge: St Malo Press.

Dretke, J. P. (1968), 'The Muslim Community in Accra (An Historical Survey)', M.A. Thesis: University of Ghana.

Esposito, J. L. (ed.) (2003), *Oxford Dictionary of Islam*, Oxford: Oxford University Press.

Harrell-Bond, B., Howard, A. & Skinner, D. (1978), *Community Leadership and the Transformation of Freetown (1801–1976)*, The Hague: Mouton Publishers.

Janson, M. (2005), 'Roaming About for God's Sake: The Upsurge of the *Tabligh Jama`at* in the Gambia', *Journal of Religion in Africa*, 35 (4), pp. 450–81.

Linden, I. (2004), 'Christianity, Islam and Poverty Reduction in Africa', London: SOAS (unpublished presentation).

Skinner, D. (1976), 'Islam and Education in the Colony and Hinterland of Sierra Leone, 1750–1914', *Canadian Journal of African Studies*, 10 (3), pp. 499–520.

Skinner, D. (1983), 'Islamic Education and Missionary Work in the Gambia, Ghana and Sierra Leone During the 20th Century', *Bulletin on Islam and Christian-Muslim Relations in Africa*, 1 (4), pp. 5–24.

Skinner, D. (2009), 'The Incorporation of Muslim Elites into the Colonial Administrative Systems of Sierra Leone, The Gambia and the Gold Coast', *Journal of Muslim Minority Affairs*, 29 (1), pp. 91–108.

Verlet, M. (2005), *Grandir à Nima (Ghana): Les figures du travail dans un faubourg populaire d'Accra*, Paris: IRD Èditions et Karthala.

Weiss, H. (2007), *Begging and Almsgiving in Ghana: Muslim Positions Towards Poverty and Distress*, Research Report no. 133, Uppsala: Nordiska Afrikainstitutet.

Interviews Held by

Sierra Leone

Dr Idriss Alami, Freetown, 24 November 1988.
Sheikh Mohammad Ramadan Bah, Freetown, 3 June 1987.
Alhaji Sajalieu Bah, Freetown, 3 December 1988.
Hassan Bangura, Freetown, 24 and 29 August 1985, 31 May and 4 June 1987, and 20 and 22 November 1988.
Al-Hajj Y. A. T. Bangura, Freetown, 6 December 1988.
Alhaji Sanu Barry, Freetown, 3 December 1988.
Shaikh Houssein Chahade, Juba, June 1987.
Al-Hajj M. M. Fofanna, Freetown, 2–3 and 5 June 1987 and 4–5 December 1988.
Shaikh Muhammad Ridda Irawani, Freetown, 3 December 1988.
Al-Haji U. N. S. Jah, Freetown, 29 November and 3 December 1988.
Muhammad Abu Bakar Jallo, Freetown, 4 December 1988.
Alhaji Mohammad Jallo, Freetown, 3 December 1988.
Al-Hajj Musa Kabia, Minister of Rural Development, Social Services and Youth, Freetown, 6 December 1988.
Bai Bureh II Multi Mohammad Kamara, Freetown, 8 December 1988.
Dr A. B. Kargbo, Freetown, 29 November and 3 December 1988.
Bilal Kargbo, Freetown, 31 May 1987.
Al-Hajj Abdul Qadri Koroma, Freetown, 30 November 1988.
Haja Aisha Sasso, Freetown, 6 December 1988.
Shaikh Jibril Sesay, Freetown, 28 August 1985 and 4 June 1987.
Al-Hajj Kekura Allieu Deen Swarry, Freetown, 30 November 1988.
Shaikh Ahmad Tijan-Sillah, Freetown, 1 June 1987 and 26 November 1988.
Shaikh Bashar Sankoh Yilla, Freetown, 4 December 1988.

The Gambia

Muftah M. Abu Aisha, Banjul, 15 August 1985.
Al-Hajj Bashiru Darboe, Banjul, 17 August 1985 and 16 August 1988.
Al-Hajj Ahmadou C. Drammeh, Banjul, 27 June 1982 and 14 November 1988.
Ibraima Jallo, Banjul, 14 and 18 November 1988.
Al-Hajj Imam Abdullai Jobe, Banjul, 18 November 1988.
al-Hajj Shaikh Joof, Banjul, 16 November 1988.
Ousman Ndow, Banjul, 17 November 1988.

Ghana

Musa ben Ahmed, Accra, 25 November 1980.
Abdu Salaam Bansi, Nima, 3 October 1989.
Al-Hajj Abdullai Showumi Classpeter-Williams, Accra, 27 November 1980.
Al-Hajj Abdu Rahman Collison-Kofi, Accra, 26 November and 1 December 1980.
Al-Hajj Ibrahim Darpoh, Accra, 26 November and 1 December 1980, 30 March 1982 and 5 October 1989.
Al-Hajj Ahmed Futa, Accra, 2 December 1980.
Sheikh Umar Ibrahim Imam, Nima, 26 September 1989.
Ahmad Kotey, Nima, 26 September 1989.

Imam Muhammad Okine, Mabrouk, 16 April 1982.
Al-Hajj Dauda Otoo, Accra, 26 November 1980.
Sheikh S. M. T. Tabataba'i, Accra, 3 October 1989.
Armiyaou Brimah Yaya, Nima, 1 December 1980 and 1 April 1982.
Mallam Hassan Zacharia, Nima, 28 September 1989.
Mallam Husseini Zacharia, Nima, 30 September and 4 October 1989.

6
Faith-based Organizations, the State and Politics in Tanzania

Ernest T. Mallya

Faith-based organizations (FBOs) are central actors within broader civil society in developing countries, especially in Africa. From a certain perspective, one might argue that FBOs have long pre-dated other civil society organizations (CSOs) in that they either appeared with the evangelization of African societies in the nineteenth century or the Islamization of the East Coast of the continent much earlier. Civil society, of course, can be defined very broadly as the space in which self-organizing and relatively autonomous groups, movements and individuals attempt to articulate values, to create associations and solidarities, and to advance interests and occupy the space between an individual and the government (Linz & Stepan, 1996). In this 'broad church', there are, among others, voluntary groups that may be very well organized as well as those not so well organized; and there may be CSOs which assist members to interact in a manner that is beneficial in different ways – politically, socially, economically – and so on. CSOs might generally be categorized as formal or informal, or as organized or more loosely organized. The former would include such organizations as labour unions, which adhere to codified rules, may formally represent certain groups and need governmental sanctions to operate, among other conditions. Informal organizations may consist of groups of individuals who co-operate in different ways for the benefit of their own communities: for collective action, financing, and the provision of services (e.g. neighbourhood vigilante groups, user groups, and informal support groups such as burial solidarity groups). Many FBOs in Tanzania may be categorized as organized according to the criteria above. There are, however, significant numbers of more community-based FBOs, which operate at a lower level and at a smaller scale, and also differ in the amount of resources – human, financial and material – with which they operate.

This chapter will focus on the analysis of the synergy that has been created between FBOs and the government and its agents, given the fact that most FBOs in Tanzania have a development agenda similar, or at least parallel, to that of the state. It will be argued that while the state sets the rules (including those banning civil society organizations from engaging in politics) it may not have the resources to sustain its legitimacy, mainly due to its failure to provide critical social services such as education, water and health. Ironically, the sacred may end up helping the secular and legitimize its stay in power by doing what the state should have been doing. In Tanzania, while traditional CSOs like co-operatives and trade unions lost power and influence in the privatized economic domain (see for example, Ellis, 1983), the FBOs have tended to centralize and gain more power after liberal political and economic reforms. This can raise a dilemma, as when FBOs collaborate with authoritarian and corrupt governments. Civil society members and activists may not sanction close relationships between their organizations and the state on those grounds. Being too close to government can lead to agendas changing, being muzzled, losing community trust and even banning. However, in trying to influence governmental action and policy, FBOs, as well as other civil society organizations, have had to engage with the state in different ways including entering into formal partnerships with governments in order to develop policy as well as to help implement such policies. In contrast, some civil society organizations have remained distant and have chosen to act in a more traditional manner by influencing policy through advocacy strategies or outright opposition. These relational issues are the focus of this chapter.

Background to 'Religious' Tanzania

In Tanzania there are three main religious followings: Islam, Christianity and the so-called Indigenous Religions. The breakdown by numbers is hard to discern because since the 1967 census, religious affiliation has not appeared on the census form. However, reasonable estimates might be that Christians and Muslims account for two-thirds of the population. Indigenous religions, other religions and atheists make up the remainder. It is also estimated that there are slightly more Muslims than there are Christians. When talking about religions in Tanzania, people refer mostly to Islam and Christianity. Islam is said to have come to the East African coast around 700 AD when Arab traders first arrived. Since the Arabs came as merchants, the spread of Islam went hand in hand with trading. It has to be noted that in East Africa Islam was not spread via *jihad* (holy war) as in some other parts of the world. Since the Arabs' contact was initially with the coastal areas, these areas have become the stronghold for the religion, followed by those inland trade-route areas which interacted with the Arab traders. Compared to the forms of education that Arab merchants found in the areas they ventured into, Islamic education was much more formal in that it was taught in specific venues, there were formally appointed teachers, and a standard duration of learning (Mushi, 2006, pp. 418–19).

Christianity, on the other hand, came with European missionaries. The first to arrive, in Zanzibar in 1863, were of the Holy Ghost denomination. They moved to the mainland in 1868, settling at Bagamoyo (Mallya, 1998, p. 241). By the time Germany annexed Tanganyika, there were already 5 missionary groups operating there. With the multiplication of these religious groups, some missionaries moved to other parts of the country such as Kilimanjaro. The European missionaries respected local languages and in most cases they imparted education (both religious and secular) in those languages. Local communities were often suspicious of the missionaries but with the onset of colonial rule (and with the penetration of the cash economy posing new pressures and opportunities), parents willingly sent their children to mission schools. By the time the First World War broke out mission schools had enrolled more pupils than schools run by the colonial administration. The difference in the content of what was taught by the two religions was to have some impact in the future of Tanganyika. Muslims were taught religious matters and the Arab language with minimal secular education whereas Christians combined both religious and secular education. Mission schools were exclusively for those who converted to Christianity.

In the history of Europe, it has been shown that religion and state were closely related for a long time. In some states this relationship is still very vibrant; in some, those who hold religious authority have political authority as well (Moyo, 1982, p. 67). During this era, religion in many European contexts was used to, among other things, legitimize state authority (Smith, 1971, p. 2). In Africa today there are two main state-religion relationship categories – the confessional states and the secular states (Moyo, 1982, p. 63). The former category includes those countries that have declared that religion has a lot of input into political processes, for example, Sudan and Libya. The latter category includes those states which claim to be secular; in other words they allow religious freedom for their citizens but formally separate religion with politics. Tanzania is constitutionally such a state.

While the constitution stipulates that the state is secular, the situation on the ground presents a different picture. Even when one traces secularism in development theory the same would appear to be the case – that development in Tanzania would barely appear as secular development.[1] Secularism holds that everything is worldly and rational, while scientifically unfounded beliefs, superstition etc. should be put aside if man is to act freely. But to be secular, it would seem that the other characteristics of a modern society ought to be present. Within this Western, secular approach, science liberates individuals from superstitious beliefs that one would find haunting individuals in underdeveloped societies.

When one compares the reality of Tanzania as far as other characteristics are concerned, doubts regarding Tanzania's secular, modern image emerge. The population is, by and large, of a more collectivist character than individualist. Indeed, the policies that were followed until recently encouraged the population to have that character. Traditionalism, which is still dominant in the rural areas, encourages collectivism rather than the individualist pursuit of advancement. The application of scientific knowledge and technology is at a very low level. The rural population which comprises more than 70 per cent of Tanzanians use low-technology tools for their productive activities such as the outdated hand hoes. Tanzania is just beginning the process of institutionalizing the brand of democracy that would allow for a representative, liberal government. The multiparty system introduced in 1992 is still shaky, with most of the parties finding it difficult to contest elections. As for income, Tanzanians accrue some of the lowest per capita incomes in the world. According to the Minister for Finance and Economic Affairs in his 2009 state-of-the-economy speech, Tanzania's PCI in 2008 was US$525 (URT, 2009). In developed countries, PCI is as high as US$55,000 (www.IndexMundi.com). For most of the 1980s Tanzania's PCI was falling rather than growing, and only started to pick up again in the mid-1990s. When one looks at rationality which implies tiers of standardization, consistency and order that each and everyone ought to abide by, it is all too clear that rules and regulations are not followed in government bureaucracy, nor are they followed in many other sectors of society. High levels of corruption in public service are indicative of this. Lastly, urbanization in Tanzania is also at a very low level, in that only about 25 per cent of the population is urban-based compared to the developed world, where the average is as high as 73 per cent. Although secularization has been institutionalized, the process of separation between the State and the Church in public matters is still questionable, and this accusation has been verbalized by many in Tanzania, especially by Muslim voices (see for example, Jumbe, 1994).

When it comes to service delivery, the state in Tanzania does not hide its need for assistance from religious organizations in development issues, as statements by various political leaders have indicated. Indeed, other indicators of this religiousness of the state include: the national anthem (with the title 'God Bless Africa') in which God is asked to do nearly everything including consolidating freedom and unity; the fact that religious leaders are invited to participate in every National Day celebration, and are asked to pray for the nation; the opening of national (state) radio broadcasts with prayers from the two major religions – Christianity and Islam; and the tendency, when natural disasters strike, for politicians to openly advise the people to seek refuge in prayers rather than science and technology.

With the state's limited capacity to implement development in the country, religious organizations are frequently called upon by the government to help in the development process. They have especially focused their efforts in education, health, utilities and other social services, areas to which the government directs a lot of its resources.

Civil society organizations, governance and the African scene

Associational life in Africa existed well before the coming of colonial powers. As for religious organizations in Tanzania, it has been noted that Arabs came centuries before Christian missionaries and they spread Islam along the coast and into a few inland regional spots like Tabora and Kigoma. Christian missionaries came at more or less the same time as the coming of colonialists (Welch., 2003). Welch, for example, talks of the founding of human rights organizations as early as 1839 to fight slavery on the continent. Civil society organizations in Tanzania date back to before the coming of colonialists (Chazan et al., 1988, p. 73). The coming of the colonial powers and the institution of the state structures we know today marked a new arrangement as far as civil society was concerned. The type of rule that existed in most of the pre-colonial societies had different rules and relationships that shaped associational life in different ways compared to those which developed after the institution of the colonial state, which came up with structures – political, economic and even social – which 'interfered' with the way life in general was conducted. This led to 'self-help' civil society organizations that catered for the economic and social adjustments needed by members who were variously affected by, among other things, urbanization, the newly monetized economy and labour migration. These organizations were the microcosms of 'professional' and trade associations that would form at a later stage, mostly in urban settings but also in plantations. Some of these associations came to take on overtly political agendas, leading to some turning into nationalist movements.

Given the co-operation that existed between the colonialists and the missionaries around the delivery of social services, a close relationship between these organizations and the state emerged. As noted earlier, the target of the two was in most cases the same – the people. Normally CSOs are organized to cater for common needs, interests, values or traditions, so the ensuing energy from their co-operation with, firstly the missionaries and colonialist authorities and later the state could and now can be channelled to a variety of activities. It has been noted that a strong and active civil society is the foundation on which rest the four pillars of governance: transparency, accountability, participation and the rule of law (ADB, 2000, p. 555). This view is supported by Mwaikusa (1996, p. 79), who emphasizes that in order to control the government, other entities beyond the state organs need to be involved in these areas. It has also been noted that in most state-civil society confrontations, the root cause happens to be the feeling that stakeholders are alienated from the decision-making process – a process which makes key decisions that affect the livelihood of the people. Conflicts arise when one religion feels that it is being sidelined in decision-making especially with regard to resource allocation (Tambila, 2006, p. 57). Further, the developments in both the economic and political spheres in the last two decades of the twentieth century created even more need for civil society organizations as demand for their presence increased. In Tanzania FBOs have tended to be confident when demanding 'favours' such as economic and political support from the state because leaders have generally been part of one or other religion.

There was increased FBO activity in Tanzania in the years when there was serious economic decline (Wagao, 1993). This increase can be linked to several dynamics. First, it has to do with reduced state capacity due to diminishing access to resources. When the state failed to deliver, people tried to co-operate in order to make life possible under the circumstances, and religion and ethnicity often happened to be the axis around which organizations developed. Secondly, there was more activism, given the developments in the area of civic and political rights as demanded in other parts of the globe. Thirdly, the processes of liberalization opened the floodgates for CSOs to form. For the African continent, the degeneration of state institutions due to poor economic performance from the mid-1970s through to the 1990s led to the erosion of the states' capacity to deliver its classical outputs of social services – infrastructure conducive for socio-economic development, law and order and so on. In some countries the situation was so bad that the citizens saw no reason to interact with the state. They rather avoided it as it became increasingly irrelevant, as discussed by authors like Hyden (1981) and his theory of the 'Uncaptured Peasantry'. Tanzania was no exception and this phenomenon of reduced state capacity largely accounted for the resurgence of civil society organizations from the mid-1980s onwards. Statistics show that before 1990 there were very few NGOs and only FBOs were allowed to operate 'freely' in the controlled single-party political atmosphere; for some reason the political leaders tended to believe that they were not a threat to their power. However, as we shall show later in this chapter, the FBOs were delivering vital services to the population which the state did not have the capacity to do (Mukandala, 2006, p. 1).

As for the developments in the demand for political and civic rights, this was spearheaded by activists mainly targeting changes in the political system – specifically the reintroduction of multiparty politics. The shift from single to multiparty politics in Tanzania was part of the wider global picture – that of reforms towards more democratic systems in formerly dictatorial states. In a way it was inevitable. But as it came, it also came with other developments. Many political and semi-political civil society organizations were formed both before and after the reintroduction of multiparty politics; and a large proportion of these were human rights groups who demanded more space in the political arena, and a rewriting of the Tanzanian constitution (Shivji, 2003a, pp. 8–9).

The third factor – that of liberalization – has to do with the weakening of the formerly state-controlled, monopolistic NGOs, especially co-operatives and trade unions, which were 'liberalized', allowing for the formation of independent, more fragmented co-operatives and trade unions in practically all areas. Formerly, these institutions were not independent organizations as they should have been, but they were, nevertheless, fighting for their members' rights in difficult circumstances, and with some coherence as well as state support from time to time, they appeared solid and united. Liberalization weakened them in various ways – and by this process FBOs became more powerful and moved towards the centre stage in terms of court and state resources. Firstly, there was the new freedom which led to some members of existing apex organizations to withdraw and either create other umbrella organizations or go it alone. As one would expect, apex organizations in the single-party states of Africa were impositions, and members were coerced to be part of these organizations. For the state and private investors this was a welcome move because the state did not want strong unions as they might scare off investors, and investors preferred to operate in a context of a less unionized workforce. Secondly, the loss of membership led to less financial capacity as contributions dwindled. Being weaker financially led to a reduction in union activities including less mobilization in favour of the members. Thirdly, those who left the umbrella organization also had problems with regard to finances and a helpful atmosphere in which to conduct their affairs. Fourthly, the privatization of many of the parastatals led to different labour relations and regulations, some of which banned union activities at the work place. It also led to less financial contribution through retrenchment of members, due to reforms in the public and parastatal sectors. Fifthly, the private crop buyers undercut co-operatives by initially offering better prices and, once the co-operatives were out of business, lowering the prices to levels below that which had been offered by the co-operatives. The effects of liberalization on these two important organizations, co-operatives and trade unions, led to the creation of some more space for smaller, sometimes more informal, civil society organizations which began to work at the grass-roots level. This is significant because it is at the grass-roots where civil society is most readily mobilized around local issues.

Non-governmental organizations and state relations in Tanzania: A delicate coexistence

When African countries started gaining their political independence, the state was seen as the main agent of development. This was partly because dominant theories at the time – especially those from the Eastern Bloc – advocated the centrality of the state in the development process. Moreover, during the struggles for independence, nationalist leaders promised a range of economic and social changes which, given the circumstances, could only be met through the agency of the state. In most countries, there were no indigenous entrepreneurs with enough capital to take on the challenges that came with independence. The state, therefore, for a considerable period of time, became an engine for development and the provider of goods and services. In this framework of provision in the newly independent states, political leaders became patrons, creating networks of patron-client relationships, characteristic of African politics and which, to a large extent, have led to the characterization of the African state as 'patrimonial' (Sandbrook, 1993).

In post-independence, Tanzania's policy-making process was designed to take a top-down approach. Few people (those who made the Central Committee – CC – of the only party) took part to the decisional political process. Deliberate efforts were made by the ruling elite to alienate citizens from the policy process by concentrating policy-making power into fewer and fewer people. This is attested by the ruling elite's move to shift policy-making power from the parliament to the party's National Executive Committee (NEC), where the CC became the ultimate policy-maker. Citizens' alienation from the policy process took place on two fronts. First, independent civil society organizations were not allowed, and secondly, public policy-making bodies were subordinated to the party organs. Consequently, citizens lacked autonomous avenues for participation in the policy process. The adoption of the single-party constitution in 1965, and the subsequent rise of the party to the supreme organ of the state, elevated party organs to the national policy-making level. The cabinet, which should have been at the centre of the policy-making model as per the inherited Westminster model, was deprived of its power over policy-making, and was replaced in this by the CC of the party. The parliament became a committee of the party. In fact, it was subordinated to the party's National Economic Council (NEC). What needs to be noted here is that these structural changes placed policy-making power in party organs, thereby giving the party chairman – who was also the president – enormous powers (Mushi, 1981). The consolidation and institutionalization of the state and party supremacy was carried under the banner of 'nation-building', thereby justifying the exclusion of citizens and their non-governmental organizations from policy-making, as well as forcing them to be affiliates of the party.

However, the history of faith-based organizations followed a different path. The fact that these organizations were delivering many of the social services that the state was unable to offer, made it possible for them to be excluded from the process of nationalization that took place between 1961 and 1967, when the Arusha Declaration nationalizations were implemented.[2] As noted earlier, schools and hospitals owned by religious organizations were nationalized for one main reason – to make these facilities accessible to all regardless of one's religious beliefs. Even then, FBOs tended to have some freedom of their own, vis-à-vis 'secular' NGOs. The Catholic Church, for example, exercised a certain degree of power, to the extent that its advice was heeded by the government (Sivalon, 1992). Given the dependence the government had on religious organizations when it came to the provision of social services, the government had to allow them to build new facilities after the initial nationalization, and agree that they were not going to be nationalized again. When, from the mid-1970s to mid-1980s, the economy started to perform poorly, the nationalized facilities were in bad shape. The government negotiated with international financial institutions like the World Bank and the International Monetary Fund (IMF) out of which a deal was struck which demanded that the government had to withdraw from some service provision or reform their provision drastically. This led to some of the nationalized assets being returned to their previous owners. It is for this reason that the percentage of social services delivered to the communities by FBOs remained high despite nationalization.

The state in Tanzania has had different relationships with non-governmental agencies, including NGOs and other CSOs, depending on the perceived benefit and threat to itself. Traditionally, CSOs, and particularly NGOs, have seen themselves as needing to keep some distance between themselves and the state if they are to achieve their stated objectives. As Jjuuko (1996, p. 194) observes, some independent NGOs regard themselves as diametrically opposed to the state or at least as sometimes being incompatible – a stance I believe should not always be the case because there will be instances when these institutions will need to engage the state through such tactics such as lobbying and advocacy. Some NGO activities also need sanctioning by the state, a fact that makes co-operation of some kind between these actors necessary. However, as Kirsten noticed, some non-governmental organizations do not have any other choice but to engage with the state if they are to attain their goals of influencing policy and governmental action (Kirsten, 2004, p. 16).

There are three general state-CSO scenarios outlining state-CSO relationships (Kiondo, 2001). The first one is a situation where the state sees the CSOs as being its partner in development and other state activities, and therefore CSOs are seen as being supportive of the state. In most cases, these activities are the ones related to social welfare and advocacy. In Tanzania, many CSOs and, to the best of our knowledge, all FBOs fall into this category. These FBOs would include those like CARITAS Tanzania which is a Catholic operations arm and which assists people with difficulties especially when disasters strike; Tanganyika Christian Refugee Services (TCRS) which once concentrated on helping refugees but now also deals with economic empowerment programmes; BAKWATA which runs a series of social services including schools, health centres and orphanages; and Dhinureyn which is a Muslim FBO running dispensaries, schools and homes for the elderly in some regions of Tanzania. The second scenario is where the state feels that its interests are threatened by the activities of the CSOs. These activities would include situations where the CSOs were providing a political platform – whether real or imagined – for political opponents to those in power. Unfortunately, in Tanzania and elsewhere in Africa, most of the CSOs and FBOs that deal with human rights and gender sensitization happen to be seen from that angle by the state. This has been explained by the fact that while the state is supposed to be the guardian of human rights, it has often been the biggest violator of these rights. Given the low level of literacy and civic competence among Tanzanians, for example, these CSOs and FBOs have been making their members and other beneficiaries aware of their rights and urging them to demand these rights from state organs; and this is seen, in most cases, as confrontation.[3]

The third scenario is the one in which the state feels that national security is at risk due to the activities of some CSOs. In most cases, this involves the sources and levels of funding. When funding is from outside the country, and the amount is, from the standpoint of the state, too large, then questions start to be raised. Monitoring of the activities becomes closer and generally the CSO in question faces multiple legistical hurdles whenever they come into contact with the state. Again, the most likely candidates in this basket are quasi-political organizations. FBOs in Tanzania have also been outspoken when it comes to issues of human rights and resource distribution. Generally they are advocating for good governance. Furthermore, the activities in which they are involved in – education and health, for example, need large amounts of funds, most of which have came from outside the country. When it comes to mobilizing internal resources the FBOs have an advantage and are able to mobilize funds by invoking religious obligations, which has proven to be very effective. The government finds it less easy to monitor FBO funds in comparison to monitoring conventional NGO funds as the nature of the NGO Policy (and subsequent law) requires NGOs, local and international, to make available annual financial and/or audited reports to the Registrar's Office and other stakeholders (URT, 2002). The interaction between CSOs and the state in this scenario includes the recent shift on the part of donors and aid agencies to want to use CSOs as channels of development aid as well as including them as part of the policy process (Tripp, 2000; Mercer, 2003).

In practice, these three categories of the state-CSO relationships, are not always so clear-cut. There are instances where the issue of the day determines the reaction of the state. For instance, when Baraza Kuu la Waislamu Tanzania (BAKWATA)[4] runs educational institutions the state is supportive because this is in line with its own policy of promoting education. But when the same FBO initiated the demand for the reintroduction of the Kadhi Courts on Tanzania Mainland – to be financed by the government – the relationship became less cordial because the Tanzania constitution precludes the government from engaging in religious matters. The BAKWATA and other Muslim organizations threatened that Muslims were not going to vote for the party in power if the issue was not resolved as soon as possible. It is yet to be resolved.

In concluding this section, it is relevant to say that CSO-state relationships need to be looked at from the point of view of the CSOs, and it also needs to be considered whether the government is ready to participate in that relationship. In some cases there can be mutual benefit. In other cases one party may be reaping benefits that are greater than the other. It is also important to look at the nature of the state. As we have noted, the possible continuum that non-governmental organizations can experience in their relationship with the state can run from a lifeline, through a symbiotic rapport, to a possible death. If the regime in place is dictatorial, undemocratic and a violator of human rights, a close co-operation could well be 'a kiss of death' to the organization in question. The state will find ways to either emasculate it by donating 'gifts' which will muzzle the organization, or it may just legislate the NGO out of business. If the government is more democratic, observes the rule of law and needs the co-operation of NGOs and especially FBOs for the implementation of some programmes beneficial to the community, then this relationship can produce positive outcomes. A series of other relationship patterns that would, in one way or another, be of mutual benefit can result (Baer, 2008).

FBOs and service delivery in Tanzania

Despite the cautious note about the relationship between the state and CSOs in matters of politics, FBOs have a large potential if the government was to co-operate with them in the implementation of development plans; this would be a relationship that would help the government to better deliver public goods and services. This is all the more true when we consider that the mushrooming of CSOs in the aftermath of the economic crises of the 1970s and 1980s and the reforms of the 1990s was due to the virtual breakdown of the social services networks created by the nationalist governments after independence, and supported by the older CSOs, specifically FBOs. CSOs can help government deliver services more efficiently and effectively through the identification of target groups, facilitating their access to the services on offer, and even coordinating the delivery of services from various sources. Long before the present government came into power, FBOs have often been working in particular communities. They have experience, networks and personnel to deliver services in the country. In developing countries like Tanzania where infrastructures are poor, FBOs have been used as the most reliable and effective vehicles for service delivery where governments failed. FBOs in particular are known to have reached people that governments have failed to reach, especially where roads, telephones and the like are acutely inadequate; or where the government would not venture because of some political reason. And, most importantly, FBOs have the potential to provide checks and balances on abuses of power at different levels of the 'implementation structure'.

In Tanzania, FBOs have played a big role in health and education since the pre-independence era. They are now very active in service sectors as well. Lange et al. (2000) for example, note that in 1986 the state called on churches and other NGOs to play a greater role in the provision of health and education services. They note, for example that, in the education sector, non-governmental organizations were running 61 per cent of secondary schools, 87 per cent of nursery schools and 43 per cent of hospitals in nine districts in Tanzania in 1993. Similar percentages would be true in many other districts. Today it is the schools owned by the religious organizations that perform best in examination (URT, 2008).[5] It has now become common to hear that the best ten or so schools are seminaries and schools run by FBOs. This has an implication even for the future in that these graduates will make the pool from which future leaders will be drawn from as they are likely to excel to the top level. This is no small contribution to a country like Tanzania. When it comes to university education, four of the twelve universities in Tanzania are owned by religious organizations. These organizations have other tertiary institutions as well like teachers' colleges and similar education facilities. In the health sector FBOs have had a big role in Tanzania as Table 6.1 below shows.

Statistically, FBOs own 25 per cent of referral hospitals and 37 per cent of the district level hospitals, which are two critical levels in the health system. People tend to go to these hospitals because of the greater

Table 6.1   Health facilities in Tanzania by ownership

Facility	Agency				
	Govt.	Parastatal	Religious	Private	Others
Consultancy/Specialized Hospitals	4	2	2	0	–
Regional Hospitals	17	0	0	0	–
District Hospitals	55	0	13	0	–
Other Hospitals	2	6	56	20	2
Health Centres	409	6	48	16	–
Dispensaries	2450	202	612	663	28
Specialized Clinics	75	0	4	22	–
Nursing Homes	0	0	0	6	–
Private Laboratories	18	3	9	184	–
Private X-Ray Units	5	3	2	16	1

Source: Ministry of Health Statistic (Website: www.moh.go.tz; last updated on 6 December 2007).

availability of drugs and other supplies at FBO centres as compared to the many dispensaries and state health centres where the government struggles to reliably provide the necessary supplies (Munishi, 2004). It is also important to note that the 'private' category came up well after 1990 when there was some relaxation of rules as, previously, medical staff working in government facilities were not allowed to operate private health facilities. This means the 'religious' category had a very big contribution prior to that policy change and many of these facilities have been there from before independence. A clear example is the conversion of some religious organizations' hospitals into Designated District Hospitals – in the Table above there are 13 of them. This means that these 13 districts are ones which the government has still not reached, although it does give such hospitals subsidies and in some cases pays the salaries of specialists and other senior medical staff transferred to these facilities. While private operators prefer to have their facilities in urban areas where they can make profits much faster, most of those of the FBOs are located in the more difficult, remote areas of the country.

Having said that, we need to remind ourselves that, the provision of services in support of the government notwithstanding, at some point in their history most if not all FBOs have also had difficulties with the state. During the socialist era in Tanzania, and more specifically during the nationalization exercise after the promulgation of the Arusha

Declaration in 1967 which sent Tanzania into socialist experimentation, many religious organizations 'lost' their properties to the state, including schools, hospitals and even land. Given the difficulties that the government has faced in running these facilities in the new neo-liberal economy, many of these properties have been returned to their original owners. However, some church schools turned government schools could not be returned given the composition of the new student community (e.g. a Christians-only school that now had a religiously mixed student community) and the investment the government had put into these facilities since their nationalization.

Another example of FBOs and other CSOs that have been co-operating with the state would include the many advocacy groups which are in the good books of the government. Examples would be those dealing with the HIV/AIDS campaigns in communities. These are doing a good job, which the government cannot do alone given the resources needed and the scope of the exercise countrywide. There are others in areas of environmental protection, civic education and good governance. In this case of co-operation one can say that the relationship has been that of a lifeline to some FBOs and some CSOs, as the government facilitated the availability of funds, as well as a symbiotic relationship in that the state has had its functions performed by these non-state actors.

The dilemma for faith-based organizations

There are issues of accountability when FBOs engage in service provision – whether funded by donors or by the state. In the case of health in Tanzania, for example, the government has been asked by FBOs to pay salaries for some staff in the Designated District Hospitals. This aspect has two sides to it. First, would the FBOs' staff still be accountable to their members/employers? Would they still be pursuing their original goals? In health the original goal could be very different from what they are doing now – as would be the case of designated infrastructures. How much will they be able to do in order to balance the forces – the need to perform the core activities for which they were formed, and the need to have the state on their side in order to have access to resources,? Secondly, there is the government's accountability to the electorate and the donor as far as the use of resources is concerned. FBOs are not traditionally vehicles for the delivery of services paid for by taxpayers that need to be accounted for in parliament or lender/donor funds that need to be repaid. When something goes wrong there can be problems for the government in accounting for its decisions, although substantial arm-twisting might have come from the donors forcing these actors be part of policy implementation.

Where FBOs and other CSOs have been co-opted into service provision there can be other repercussions. As noted earlier, by helping the state to deliver services, whether the services have some donor funding as a way to avoid corrupt and undemocratic governments, or are provided by an FBO doing what it believes it should do as a faith-based organization, the state may end up being the net beneficiary, depending on its history of course. While donors may feel that they are rebuking corrupt, undemocratic and inefficient regimes by channelling aid money through non-governmental organizations, this could be a way to cleanse the state in the eyes of the electorate. As Whitefield (2003, p. 383) remarks,

> The powerful influence of donor agencies on local organisations, combined with the application of civil society as an idea to achieve their objectives of economic and political liberalisation, may work towards stabilising the existing social and political order.

Depending on the timing of the activities, which generally rise up when there are harsh conditions, the performance of the FBOs may raise expectation and confidence in the regime thereby leading to the regime staying in power longer than it should have, if the current conditions had not been in place. Under these negative conditions the state may even channel resources to friends and foes alike, a thing that would be impossible were the state to be the direct provider of the services. In short, FBO activities of that nature can very well prop up otherwise unwanted political regimes, especially in countries like Tanzania where the majority of the population is not educated enough to make informed choices when it comes to electing government leaders.

With the democratization process taking place in many countries in Africa, churches and mosques have tried to affect government's action and policies. In Tanzania churches and mosques have a lot of influence over the government given the facilities they provide in service to the people. While the other types of CSOs would not so openly discuss or engage in politics, FBOs have had the courage to outspokenly oppose the state. In 1995 religious organizations encouraged their followers to participate in the political processes going on in the country, including elections. When these first multiparty elections were due, religious organizations organized seminars and printed some materials to help their followers to understand the politics of the day. Such organizations included the Evangelical Lutheran Church of Tanzania (ELCT),

Tanzania Episcopal Conference (TEC), Christian Council of Tanzania (CCT), The National Muslim Council of Tanzania (BAKWATA) and so on. The CCT's *The CCT Position on Prevailing Situation in the Country* published in 1995 clearly states in its introduction that it aimed at 'educating our Tanzanian Society on political and socio-economic issues prevailing in the society today' (CCT, 1995). It also stated that it aimed at expressing the obligation of its member churches in dealing with and being concerned with such issues. In the document, several political issues are discussed including the relationship between the churches and the state; the church as an advocate of political justice and democracy; political and economic democracy as related to the provision of social services; and the constitution of Tanzania and the need to convene a constitutional conference.

Conclusion

It has always been a challenge for NGOs, including FBOs, to define their level of collaboration with the state. FBOs have been involved in the same agenda as that of the government for so long and with such intensity that this relationship has cast doubts on the secular status of Tanzania. Given their deep involvement, FBOs also have a substantial say in some policies as well as in setting priorities in the broader policy process of the government. In the process, as has been mentioned, there are collaborative efforts in which such items as funding are organized jointly. FBOs are not in a position not to co-operate with the government as their members would suffer without their ability and plans to help alleviate difficulties in day-to-day life. Some freedom is definitely lost here. However, religious organizations should not automatically assume that the state and its apparatuses are a threat to them and their independence in particular. Given that their followers' interests are paramount, any coalitions, partnerships or co-operation that will ensure that this is attained should be taken seriously.

To conclude, it is not being suggested that CSOs, including FBOs, should take over the role of representing the people in policy-making, especially in Tanzania where many of those that can engage in policy dialogue are elite NGOs, as clearly noted by Mercer (2003, p. 749) and scathingly attacked by Shivji (2003b, pp. 2–3) when he remarks that NGOs have begun to 'lose their sight' about what they should be, what they should be doing and what they should not be doing. He terms all this 'participation by substitution' implying that NGOs are alienating core stakeholders in the policy process rather than helping the 'people' participate in the policy process. Contrary to Shivji's thesis, FBOs seem to qualify somewhat as representatives of the people in that they are not selective – as what gets people under their umbrella is faith and not variables that create elites and non-elites. FBOs are known to have their focus on the poor and vulnerable, working in remote areas of poor countries where government is yet to 'reach'. Other NGOs tend to work from the other end – as both urban-based and single-issue based, and requiring the services of the educated. FBOs, on the other hand, are all-inclusive and operate in both rural and urban areas.[6] This is reiterated by a report to the US Congress on progress made with regard to the US President's Emergency Plan for Aids Relief (PEPFAR) in Africa (including Tanzania) which states:

> Local community- and faith-based organizations remain an under-utilized resource for expanding the reach of quality services. They are among the first responders to community needs, with a reach that enables them to deliver effective services for hard-to-reach or underserved populations, such as people living with HIV/AIDS and orphans.
>
> (US State Department, 2005)

In this era of public sector reforms, where the private sector has been identified as the number one government partner and has a variety of co-operation arrangements in place, the private sector has come to see the NGO sector as a threat in that both are sometimes vying for the same resources. FBOs, however, have an advantage over the private sector most of the time as they do not work for profit even when their efficiency levels are the same. If the two operate on a level playing field, the FBOs will almost always deliver better results, as noted by the above-quoted PEPFAR report.

The challenge is, therefore, for the government to see how these two partners in development can be harmonized for the betterment of the lives of the citizens. Meanwhile, the Tanzanian FBOs remain powerful when it comes to many issues due to the centrality of their activities for the government of the day, and this is likely to be the case for the foreseeable future.

Notes

1. For the characteristics of a 'modern' society see, for example, modernization theorists like Levy (1966) and Black (1966).
2. For instance, the government depends a lot on FBOs to start and run orphanages and homes for the elderly, and it has very few kindergartens compared

to those run by FBOs. Today the needs have grown exponentially given the HIV/AIDS pandemic.

3. A good example is HAKIELIMU which encouraged tax payers to make further money available to education so that children would get better education. However, the government ended up banning HAKIELIMU's value-for-money campaigns, saying that it was meddling in government business.

4. This Swahili name translates into the National Muslim Council of Tanzania.

5. The document states that 'regarding to the performance of Division I–III candidates from Seminary Schools performed better, followed by Government and Non-Government Schools. While Community Schools experienced the worst performance over the period under SEDP, 2004–2008'.

6. Examples of these would include Adventist Development and Relief Agency (ADRA) of the Seventh Adventist Church which works in the rural Hanang District in Tanzania which is remote, hard to access and also inhabited by the few remaining hunter-gatherers by the name of Hadzabe. ADRA is also in Kigoma Region which is the hardest region to access from Dar es Salaam, the business centre for Tanzania.

References

Asian Development Bank (ADB) (2000), 'Civil Society and Non-Governmental Organizations' in ADB, *To Serve and to Preserve*, Hanoi: ADB.

Baer, F. (2008), 'FBO Health Networks and Renewing Primary Health Care', Consultant for Health Systems Development, World Bank.

Black, C. E. (1966), *The Dynamics of Modernization*, New York: Harper and Row.

Christian Council of Tanzania (CCT) (1995), *The CCT Position on Prevailing Situation in the Country*, Dodoma: CCT.

Chazan, N., Mortimer, R., Ravenhil, J. & Rothschild, D. (1988), *Politics and Society in Contemporary Africa*, London: Macmillan.

Ellis, F. (1983), 'Agricultural Marketing and Peasant-State Transfers in Tanzania', *Journal of Peasant Studies*, 10 (4), pp. 214–42.

Ellis, F. (1988), 'Tanzania' in Harvey, C. (ed.), *Agricultural Pricing Politics in Africa, Four Case Studies*, Basingstoke: Macmillan.

Hyden, G. (1980), *Beyond Ujamaa in Tanzania: Underdevelopment and an Uncaptured Peasantry*, Berkeley: University of California.

Jjuuko, F. W. (1996), 'Political Parties, NGOs and Civil Society in Uganda' in K. K. Oloka-Onyango & C. M. Peter (eds), *Law and the Struggle for Democracy in East Africa*, Nairobi: Claripress, 180–198.

Jumbe, A. (1994), *The Partnership: Tanganyika-Zanzibar Union – 30 Turbulent Years*, Dar es Salaam: Amana.

Kiondo, A. (2001), 'Civil Society: The Origin and Historical Development of the Concept in Tanzania'. A paper presented at a workshop under the Johns Hopkins Comparative Non-Profit Sector Study, Bagamoyo, 30 November–1 December.

Kirsten, A. (2004), *The Role of Social Movements in Gun Control: An International Comparison Between South Africa, Brazil and Australia*, Research Report No. 21, Durban: Center for Civil Society Research.

Lange, S., Hege, W. & Kiondo, A. (2000), *Civil Society in Tanzania*, Bergen: Michelsen Institute.

Levy, M. (1966), *Modernization and the Structure of Society*, Princeton: Princeton University Press.

Linz, J. & Stepan, A. (1996), 'Towards Consolidated Democracies', *Journal of Democracy*, 7, pp. 14–34.

Mallya, E. T. (1998), 'Group Difference and Political Orientation: Religion and Education' in S. S. Mushi, R. S. Mukandala & M. L. Baregu (eds), *Tanzania Political Culture: A Baseline Survey*, Dar es Salaam: Interpress.

Mercer, C. (2003), 'Performing Partnership: Civil Society and the Illusion of Good Governance in Tanzania', *Political Geography*, 22, pp. 741–63.

Moyo, A. M. (1982), 'Religion and Politics in Zimbabwe' in P. Kirsten (ed.), *Religion, Development and African Identity*, Uppsala: SIAS.

Mukandala, R. S. (2006), 'Introduction' in R. S. Mukandala, S. Y. Othman, S. S. Mushi & N. Ndumbaro (eds), *Justice, Rights and Worship: Religion and Politics in Tanzania*, Dar es Salaam: E&D.

Munishi, G. K. (2004), 'Quality Health Care in Tanzania: Emerging Issues' in G. Mwabu, G. Wang'ombe, D. Okello & G. Munishi (eds), *Improving Health Policy in Africa*, Nairobi: University of Nairobi Press.

Mushi, P. A. K. (2006),'Religion and Provision of Education and Employment in Tanzania' in R. S. Mukandala, S. Y. Othman, S. S. Mushi & L. Ndumbaro (eds), *Justice, Rights and Worship: Religion and Politics in Tanzania*, Dar es Salaam: E&D, pp. 431–56.

Mushi, S. S. (1981), 'The Making of Foreign Policy in Tanzania' in S. S. Mushi & K. Mathews (eds), *Foreign Policy of Tanzania 1961–1981: A Reader*, Dar es Salaam: Tanzania Publishing House.

Mwaikusa, J. T. (1996), 'Party Systems and the Control of Government Powers: Past Experiences, Future Prospects' in Oloka-Onyago, K. Kibwana & C. M. Peter (eds), *Law and the Struggle for Democracy in East Africa*, Nairobi: Claripress.

Sandbrook, R. (1993), *The Politics of Africa's Stagnation*, New York: Cambridge University Press.

Shivji, I. G. (2003a), 'Reflection on NGOs in Tanzania: What We Are, What We Are Not and What We Ought to Be?' HakiElimu Working Papers, Dar es Salaam: Hakielimu.

Shivji, I. G. (2003b) 'Democracy and Constitutionalism in East Africa: Taking Critical Stock' in I. G. Shivji (ed.), *Constitutional Development in East Africa*, Kampala: Kituo Cha Katiba.

Sivalon, J. (1992), *Kanisa Katolikina Siasa ya Tanzania 1953–1985*, Peramiho: Benedictine Publications.

Smith, D. E. (1971), *Religion, Politics and Social Change in the Third World*, New York: Free Press.

Tambila, K. I. (2006), 'Inter-religious Relations in Tanzania' in R. S. Mukandala, S. Y. Othman, S. S. Mushi & L. Ndumbaro (eds), *Justice, Rights and Worship: Religion and Politics in Tanzania*, Dar es Salaam: E&D.

Tanganyika Christian Refugee Services (TCRS) (2004), *Country Strategy Outline 2004–2008*, Arusha: Evangelical Lutheran Church in Tanzania.

Tripp, A. M. (2000), 'Political Reform in Tanzania: The Struggle for Associational Autonomy', *Comparative Politics*, 32 (2), pp. 191–214.

United Republic of Tanzania (URT) (2002), *The NGOs Policy of 2001*, Dar es Salaam: Prime Minister's Office.

United Republic of Tanzania (URT) (2008), *Basic Statistics in Education in Tanzania (BEST)*, Dar es Salaam, Ministry of Education and Vocational Training; also available at www.moe.go.tz/statistics.html.

United Republic of Tanzania (URT) (2009), *Economic Survey 2009* (Dar es Salaam: Ministry of Finance and Economic Affairs).

US State Department (2005) 'Community and Faith-Based Organizations' Report to Congress on Progress of the Presidential Emergency Plan for AIDS Relief (PEPFAR), September.

Wagao, J. I. (1993), 'Religion, Policy Reform and Crisis: Tanzania' in Adenaur Stiftung, K. (ed.), *Islam in Africa South of the Sahara*, Conference Proceedings, Sankt Augustin, Germany.

Welch, C. E. Jr. (2003), 'Human Rights NGOs and the Rule of Law in Africa', *Journal of Human Rights*, 2 (3), pp. 315–27.

Whitefield, L. (2003), 'Civil Society as Idea and Civil Society as Process: The Case of Ghana', *Oxford Development Studies*, 31 (3), pp. 379–96.

7
Burying Life: Pentecostal Religion and Development in Urban Mozambique

Linda van de Kamp

Introduction

Currently, the so-called secularization thesis is the subject of much debate (Asad, 2003; Taylor, 2007; de Vries, 2008). A new paradigm is emerging pointing at the ongoing influence or even resurgence of religion in modern societies. Increasingly, it is being acknowledged that religion is central to the lives of people and that religion cannot be separated from the political and the economic. This correlates with the increasing interest in the potential of religion for development programmes (Belshaw et al., 2001; Marshall and Keough, 2004). Yet, the new openness towards the potential of religion in developing society seems to imply both a Weberean and Durkheimean legacy; namely that religion helps to generate a capitalist society, brings progress, is supportive and contributes to security, democracy, certainty and continuity (see e.g. CDE, 2008; Ranger, 2008; Berger, 2009). The question is, however, whether the realization that religion is an intrinsic part of society and people's everyday life can also be discerned in an investigation into what kind of 'development model' religions produce, which may not necessarily be directed to progress in a modernist sense. This observation arises from research carried out about Pentecostalism in Mozambique, which shows that Pentecostalism has become popular precisely because it questions and even destroys existing models of development.[1]

The recent and growing form of Christianity in Africa, Pentecostalism, is known for its 'prosperity theology' or 'health and wealth gospel' (Gifford, 2004). It underlines that a militant and courageous faith brings happiness, health and prosperity in all domains of life. A famous church in this regard is the Universal Church of the Kingdom of God (Universal

Church). This Brazilian Church is amazingly popular in the urban areas of Mozambique, where it attracts the (new) middle classes (Cruz e Silva, 2003; Freston, 2005). The Church propagates an optimistic and activist attitude in showing every visitor that 'you can make it'. The pastors encourage people to start businesses and to fulfil their dreams of a big house, a car and a happy family life. It is for this reason that some stress the hopefulness and personal agency these type of churches create (Martin, 2002), while others have found these churches to be outlets for a productive working life, business activities (CDE, 2008) and improved family life (e.g. Brusco, 1995). However, this research indicates that despite the emphasis on prosperity, only a few converts succeeded in enriching their lives economically and socially. Converts' deliberate participation in Pentecostal churches sometimes resulted in even more insecurity and poverty.

The clearest example in the Universal Church is the central ritual practice of excessive offerings of money, sometimes of several thousands of dollars.[2] The emphasis on prosperity includes the practice of 'sowing and reaping', meaning that who sows will reap, which is understood materially through financial gain – the more money one offers the more money one will gain in the future. The majority of the converts interviewed gave their money away to the church, meaning that in various cases their businesses went bankrupt. Moreover, the money appeared to be accumulated by the leadership and not invested in sustainable social welfare programmes or economic initiatives. Therefore, as in Brazil (Mariano, 2003, pp. 51–5), the critics in Mozambique accused the Universal Church of being a money-machine and a business instead of a church, and of becoming rich at the expense of the poor.[3] To explain this seemingly illogical development that people gave without positive result, Comaroff and Comaroff (2000) pointed to the similarities of prosperity gospels with occult practices and so-called pyramid schemes. While many people can only gaze at the consumer items shown in the media and the shops, only some will actually succeed in becoming rich. People wonder why their neighbour is successful while they stay poor. Apparently, capitalism consists of a magical power of making money out of nothing. Thus people may decide to give their money to Pentecostal churches to create a chance to gain wealth and at the same time maybe get rid of the dangerous powers commodities harbour (see also Meyer, 1999).

This may not be a satisfactory explanation. In Mozambique, the Pentecostals did indeed considered the capitalist economy as an enchanted enterprise. However, many did not find the neo-liberal order mysterious and dangerous. They partook in the capitalist economy. They worked in companies, banks, NGOs and ran (small) shops. They knew how to profit from this system. Why then did converts offer their money excessively to churches, when an outsider would come to the conclusion that they would do better to invest the same money in the local business they were making an offering for? The aim of this chapter is to explore the relation between Pentecostal religion and development in Maputo, the capital city in Southern Mozambique,[4] by looking at the meaning and significance for converts of *sacrificar* (sacrificing) money to the Universal Church.

This chapter explores how the traditionally established links between religion and development in Mozambique – as sustaining the circle of relations between ancestors, kin, money and fertility to guarantee the reproduction of life (i.e. development) – has evolved into particular and often problematic dimensions in the urban domain in which the Pentecostals are involved. A principal point of contestation is the role of money in shaping relations as the basis for the development of society. Upcoming middle classes, for example, experience the reciprocal kin relations as a burden because of the obligation of sharing money. In particular the new economic roles of women in the urban domain bring along complications in intimate and kin relations. Their involvement in the money practices of the Universal Church is an attempt to change a socio-economic order that is being experienced as a burden by destroying and burying that order's fundamental structures, even though the primary goal is to become born again as a Pentecostal and thereby die to the old life. The effect is increasing uncertainty and risk – thus dangers are not diminished – yet this does not divert the converts and they continue to deliberately engage in this process. This has implications for the way in which the relation between religion and development are perceived.

Burying life

In December 2006, a news item was broadcast on STV, the most popular and critical television channel in Mozambique, about a riot against the Universal Church in one of Maputo's neighbourhoods. One of the participants of this riot explained what happened.[5] In the grounds of the Universal Church in that neighbourhood people saw coffins and holes waiting to receive the coffins. Neighbours were shocked and started to throw stones at the church and the pastor until the police arrived and took the pastor away to protect him. As one interviewee stated 'it is very un-African what was happening in this Church'. People were afraid that people were being buried, which in such an inappropriate place could cause a lot of harm. Some understood it as a witchcraft ritual. The interviewee further stated that people in the neighbourhood, especially the older ones, shared the long-held opinion that the Universal Church should be prohibited. 'These older people still live like in the rural areas. Here in the city there are many other influences, people are losing their culture. The Universal Church behaves against the principles of people from here.'

A couple of times I witnessed the use of a coffin in the services of the Universal Church. The following is an extract from fieldwork notes about a Monday evening service for success in business on 14 February 2005:

> At the entrance of the Church building assistants distributed small papers on which to write down financial problems. The pastor summoned people to throw the written paper in the coffin displayed in the front of the Church. Later on, the pastor would bury the coffin. Thus our problems would go with it. The pastor reads Psalms 112:3: 'His family will be wealthy and rich, and he will be prosperous for ever'. During the sermon all present, about 1000 persons, had to speak out loudly: 'belonging to God brings prosperity'. Then, we had to fix our eyes on the coffin. The coffin was closed and wrapped up in a red cloth. The pastor said that many people are victims of witchcraft. If we had financial problems and our business was not prospering then someone sent us evil. He ordered the evil spirits to leave us all. Some people started to cry and fall down and spirits were expelled from their bodies. Everybody with problems was called to take the red cloth in their hands and while looking the pastor in the eyes he sent the evil spirit away. After that, all those remaining could come forward and touch the red cloth. First, those who gave 500 dollars could do so, then those who gave 300 and then 100, etc. Finally the red cloth was put in the coffin.

In many respects the Universal Church was regarded as distant from society by critics, such as the older people mentioned above. The pastors had large houses and large cars but they did not share their wealth with others. Stories circulated about family members who lost everything because of a relative who participated in the Church's offerings and sold a television, radio or computer. Suspicion increased because Church members had to act in secrecy. The pastors obliged converts to keep things between them and God. One should not talk about how much money he or she would *sacrificar*. The sacrifices happened as well as tithing and offering.[6] The Universal Church organized the Holy Fire of Israel (*Fogueira*) offerings twice a year. These were meant to enable converts to make a sacrifice of a large sum (a payment of a month or even 2 months' salary is the norm), in order to conquer all evil in one's life, thereby ensuring success in the future. During the year, additional sacrifices took place for specific personal purposes. The sacrifices were also called *holocaustos*, 'burnt offerings', a type of Old Testament offering in which the sacrifice is consumed totally by fire. References to fire and burning were central. The *Fogueira*, phrases such as 'the burning power of the Holy Spirit', the coffins and the red cloth[7] recognized the converts' involvement in a 'spiritual war'. Converts battled against 'devilish powers' such as ancestor spirits. The financial offerings were a materialization of the desire to erase the dependence on spiritual and social relations.

Fundamental to an understanding of the Pentecostal provocation of local notions of development is an appreciation of the entanglement of social relations and prosperity. Lundin (2007) has described how, in Maputo, social networks based on kinship are essential for guaranteeing a livelihood for all socio-economic classes in this urban setting (see also Costa and Rodrigues, 2007). Access to goods and services are mediated by one's social position in a network of relations that are often based on kinship. In line with Mauss' influential work *The Gift* (1969) kinship has been an essential social institution to regulate exchange relations necessary for the reproduction of biological, economic and socio-cultural life. The Southern Mozambican customary bride price *lobolo*, is an example of the relationship between economic gift-giving and kinship. It is through a gift, nowadays in the form of money, that the families and ancestors of the bride and bridegroom become interrelated. The exchange between the two families secures the social order as it guarantees the continuation of those families and, in consequence, of society (Bagnol, 2006). Bloch and Parry (1982) speak of the regeneration of life in which death, birth, marriage, sexuality and money are all bound up in one sphere. All of them are important and dependent on each other for the smooth running of social life. This can be observed in several ways in Mozambique; for example, the primary portion of a salary in a first or new job was given to kin and was partly used to offer presents to the ancestor spirits. The activities of the Universal Church go against these principles of ordering society and developing wellbeing.

Various researchers on African Pentecostalism (e.g. Meyer, 1998; Van Dijk, 1999) have shown that by converting to and giving their money to Pentecostal churches, members cannot maintain their reciprocal relations with kin and ancestor spirits. Giving the first salary to the church is a break with expressing gratitude and dependence for one's prosperity to relatives and ancestors. Moreover, what has been given in church cannot be used to meet family obligations. In this way the socioeconomically upwardly mobile person can get away from the financial burden they experience from family members who constantly appear on their door for money, hampering them in their aspirations to build a house, pay for their studies and save money for future plans. By offering this money in church they also break away from the spiritual powers of ancestor spirits that are considered evil in the Pentecostal context. It is no longer family spirits that influence their wellbeing but God. Yet few studies have explored how this discontinuity with the prominent worldview not only shapes new possibilities and meaning but is also problematic (cf. de Bruijn and van Dijk, 2009). Based on the case of the Universal Church in Mozambique, it is important to pay attention to the antagonistic and vulnerable situations it creates, and to the reasons why converts choose to face debt, conflict and uncertainty. The use of the coffin is a strong image of destruction that the neighbours understood very well and reacted to with strong feeling. Together with other practices, such as giving excessive amounts of money away, the neighbours of Church members see this as violating the reproduction of life. But for the converts, this is exactly the reason they joined the Church. Making offerings demonstrated the need converts felt to destroy the old before being able to start something new.

Money and reproduction: changing dependencies in the urban domain

Various Mozambicans are asking questions like 'where is our society heading?' After a socialist period (1974–89), a destructive civil war (1976–92), and the introduction of neo-liberal socio-economic and democratic structures (1990s), this question has become increasingly pertinent. Among both Pentecostals and non-Pentecostals, discussions on this question centred on conflictive relationships and gender roles. Uncertainties and opinions about topics such as to how to relate to kin and how a marriage should work stood out as central. These discussions often concentrated on the supposed roles of women. Interestingly, debates on how Mozambican society should develop are articulated in a gendered language. Because the majority of the converts are women and their money offerings are related to reproductive issues it is important to sketch a history of shifting balances in the various domains that pertain to reproductive relations.

The integration of groups of people into the economy of capitalism, as wage workers or entrepreneurs, has severely impacted social relations (e.g. Macamo, 2005). In the patrilineal kin system of Southern Mozambican societies the reproduction of families depended on the control by senior men of women's marriage, work and food preparation via *lobolo*. In the second half of the nineteenth century, existing power relations in Southern Mozambican societies started to shift with the labour migration of young men to South Africa (Harries, 1994). Older men gradually began to lose what control they had because their grandchildren could earn their own money and became independent of the larger family by having the possibility of paying *lobolo* themselves and establishing a nuclear family (Harries, 1994, p. 98).

Portuguese colonization adds a further dimension to this discussion of migrant labour. From 1926 onwards, the installation of the New State in Portugal introduced a programme to restore Portuguese influence to its colonies (Newitt, 1997, pp. 445–81). The decision of the colonial authorities to tax the population forced the native population to enter the labour market to earn wages (Penvenne, 1995). Work, including labour migration, became part of the colonial system. From this time on, work was no longer connected only to village and family life but also to the larger body of the state. This brought such profound shifts into the Mozambican households that breaks began to occur in the local circle of reproduction. For example, existing relations between men and women were challenged as Ronga[8] songs about the disrespect women had for their men during this period show (Sheldon, 2002, p. 57).

In the urban areas these developments had particular implications. The forced labour brought many migrant labourers to Lourenço Marques (now Maputo). This was a gendered development. The forced labour was directed towards men (Penvenne, 1995; Sheldon, 2003).[9] Those women who lived in the city were dependent on men, either through marriage and the income of their husband or through forms of prostitution. At the same time, several women came to the city on their own to start a new life and to earn money, mostly illegally (Penvenne, 1995, chapter 10). Because these urban women were predominately single or separated, the colonial administrators and both European and African males approached the many female-headed households in Lourenço Marques as a grave social and moral problem (Sheldon, 2002,

p. 66). In short, new economic and family formations emerged. From the beginning, the economic, legal and social space for women in this new urban setting was a profoundly contested one as they had far less access to wage-paying jobs.

Women gradually found more opportunities for legal work and after independence in 1975 urban centres became spaces where women began to find paid work. The postcolonial Marxist-Leninist-oriented Frelimo government encouraged women to go to school and to work for a salary. However, the simultaneous introduction of the 'socialist family' by Frelimo accentuated a division between domestic and waged work, resulting in tensions about gender roles. The 'socialist family' was a monogamous, nuclear family opposed to the polygamous and extended African family. According to Arnfred (2001, pp. 41–2) the consequence was that, economically, women in the city became much more dependent on men than they used to be in the countryside, because the household economies in the cities were based on wage labour, which the men mainly participated in. Moreover, polygamous relationships developed underground. Whereas in the rural areas several wives lived together and shared the work in the household, in the city the wives did not always know each other, and competition and jealousy was high (Arnfred, 2001, pp. 36–44). For women this continues to be a major concern and even a reason to start to frequent the Pentecostal churches today. According to various older women, the problem was that old structures were abandoned, but nothing came to replace them. This shaped feelings of uncertainty about gender roles, marriage and family life, resulting in constant struggles over proper behaviour and responsibilities (cf. Sheldon, 1996, pp. 8–9). Moreover, with economic decline aggravated by the civil war and the introduction of neo-liberal economic structures in the 1990s, a growing number of men could not earn a salary which further complicated the establishment of harmonious nuclear households.

However, these developments also increased women's public presence. Because their husbands were or became unemployed and costs of living increased, more women started to look for possibilities of earning money. In addition, most of the war refugees who fled to the city were women. For the first time, the number of women in the city exceeded that of men and the amount of female-headed households rose (Oppenheimer and Raposo, 2002, pp. 20–21). Because women could not rely on their family ties or marriage, they were forced to become financially independent, generating new possibilities of physical and social mobility (WLSAMOÇ, 2001, pp. 106–50). This process continued after the war, with the implementation of neo-liberal structures. More women entered education and started professional careers. These developments impinged upon established reproductive relations and defined current relationships.

Women complained about husbands failing to give financial support. Domestic violence increased because of frustrations, even more so in cases where women earned money while the husbands were unemployed (cf. CEA et al., 2000, pp. 64–5). Men said that it was difficult to get a wife when they had no money. The majority of the research participants said that they quarrelled about money and gender roles with their partners. The women all worked in professional careers and wanted to divide the household tasks. Not all men were necessarily opposed to cooking and cleaning, but most felt uncomfortable dealing with these tasks. Domestic roles in the home became a major source of tension when in-laws were made aware of the blurred gender roles within these young families. Mothers-in-law found the behaviour of their daughters-in-law unacceptable because a good woman should be at home, cooking and cleaning for her husband and they complained to their sons about this. Another source of friction was the extent to which the salaries of the couple should be shared with their respective families. The demands of extended families on urban couples further deepened conflicts within marriages and, as I encountered, these were experienced as a burden. Between relatives and couples distrust about improper sharing became evident. People complained about the impossibility of setting up businesses and developing their lives because they had to take care of poorer relatives. As soon as they had some money their relatives turned up to claim support. 'That is why we don't develop', one interlocutor said. 'The development agencies don't understand it. We cannot prosper because we always have to share our income.'[10]

Problematic access to money and distrust in relationships all figured in several witchcraft stories that circulated in Maputo. For example, stories were told about why there are so many beautiful, well-educated and prosperous women who do not marry. The explanation went that when they were children these women had been sacrificed to a spirit by their family in order for the spirit to provide wealth. As the women were in possession of these powers they were not allowed to marry (cf. Bähre, 2002). This was a grave allegation. It illustrated how structures of sharing wealth are under great pressure. As more and more people became rich while others became poorer, and as less and less 'normal' households could be found, people felt that patterns of exchange were no longer healthy. It was in this reality that the Universal Church's practices took place. The occurrences with the coffin touched upon an open nerve related to the whole field of the regeneration of life, embedded in historical socio-economic transformations. The answer of the Universal Church was not to call for a revitalization of kin relation for sustaining the regeneration of life and development. On the contrary, the pastors called for a definitive break with these patterns of dependencies with demanding *holocaustos*. Converts must bury and burn these ways of living. In the Pentecostal context those relations that are generally considered as the font of well-being had to pass away. This appeared to be especially appealing to upwardly mobile women.

The Universal Church and reproduction

In particular, upwardly mobile women found Pentecostalism attractive and fervently participated in these churches' big financial offers. About 75 per cent of the visitors and converts of the churches were women of various ages.[11] Their positions in the historical and contemporary urban developments sketched above, brought both new possibilities and instabilities depending on socio-economic positions, age and personal circumstances. These various conditions shaped women's involvements in church and the interaction between them and the pastors. Below, two cases are presented to demonstrate women's participation in the *sacrifícios* of the Universal Church.

Dona Silmara

Dona Silmara (57 years old) grew up in Maputo where her relatives had managed to work for the colonial authorities and the Frelimo government after independence. Like other women who were part of the relatively small Mozambican middle class, she received some schooling and married 'the European way' (in the first years after independence) i.e. she married without performing the local marriage tradition or *lobolo*, which had been prohibited by Frelimo's socialist policies. Her friend, Dona Isabel, also Pentecostal, recalled how she herself married without giving advanced notice to her family.[12] The two women were happy to escape all the interventions of kin in the marriage arrangements.[13] They immediately added that it turned out to be a deception. They were both, like many other women, victims of urban polygamy and domestic violence, and were later divorced.

For both, their position as divorced women, without a working career, was an enormous obstacle in their survival in the city in the midst of an economic crisis at the end of the 1980s. Their relative independence from their kin meant that they could not rely on their support, particularly given that without the social security of the *lobolo* there was nothing for them to rely on. No support could be expected from the state either. The fact that they had not lived up to the ideal of the 'socialist family' worsened their self-esteem. Dona Isabel and Silmara started drinking, eventually becoming alcoholics. It was in this period that the Brazilian Universal Church arrived in the city of Maputo (at the beginning of the 1990s) and both women started to attend the services where they learned that they could improve their lives; that if they wanted they could study and find a (better) job.

In 2005, Dona Silmara followed the Universal Church's Course for the 'Managers of the Nation's Wealth'. The course consisted of about 10 weekly lessons given by pastors after the evening services in different church branches. The course was meant for those converts with a business or plans to start one.[14] The central message was that people should think big. Limits only exist in people's minds. God gives the nations to his managers (Isaiah 54: 2–4; Psalms 2: 7–9) and the course participants were his managers. 'So', said the pastor, 'why would you not reach prosperity?' Themes were: how to expand your business through faith, what is a good manager, how to stay successful and how to overcome obstacles. Participants had to present their business plans to the pastors during a private appointment. Dona Silmara was very nervous and uncertain about her plan, and the pastor did indeed react negatively to her progress. She had several small businesses but they were not expanding. The pastor made it clear that she had to change her strategies.

During one session (which I attended) participants were obliged to make their commitment to become prosperous managers overt by throwing grape juice (a symbol for Jesus' blood) on the pastor and stating the amount of money they would sacrifice to reach their goal. Even before I could say to Dona Silmara that I would not join, she had already given me a sign to stay in my seat. Afterwards she said it would have been very complicated if I had involved myself. One had to be a very committed member of the special group selected by God to be his manager. It was clear to her and I that I was (still) not such a manager. It was only a year later, when she informed me about several more private meetings with the pastor and the pressure she underwent, that I realized she had also wanted to protect me, and probably herself, from the 'truth'. The pastor had regularly visited her businesses during the last couple of years as part of his counselling practice. To show her progress in business she had to submit monthly overviews of the profit she made to the pastor. Every time the pastor decided the amount of money she had to sacrifice in church, and it usually correlated with how much money she had gained. She ran into difficulties with her business because she could not invest enough capital into it and could not pay her employees. After several years of pressure-offerings, she finally went bankrupt. She was not the only one.

Women of Dona Silmara's generation who converted to Pentecostalism were very eager to become part of a new socio-economic and cultural order where it is possible to study, get a well-paid job and establish a marriage based on love. The Universal Church provided these women with a place in which they could gain some basic understanding of the economic market via business courses and learn how to become a successful woman as much in the economic as in socio-cultural sphere. They were pushed to make personal choices and to take control over their life. The breaks they had already partially undergone with various cultural practices and with relatives could finally become real with a definitive cut. Their dedication to practices such as burying coffins, sacrificing money and so on could affirm their break with local culture, their new position in life, and prove their progress and success. For these women the Pentecostal churches were not simply a place of support, but a setting 'that only helps those who are able to help themselves' (van Dijk, 2009, p. 110).

Luisa

Compared to the older generation, the younger generation (16–35 years old) was more acquainted with the challenges, uncertainties and conflicts of urban society. They had grown up in a period of war and during the introduction of a neo-liberal economy. They were better equipped to deal with the various demands put on them. For example, they negotiated with the pastors about their sacrifices more often and left church faster when their lives did not improve. At the same time, the uncertainties and risks that were part of their daily life put the practices of the Universal Church in line with 'normal' occurrences in their society.

The first time I met Luisa (27 years old) in October 2006, we talked as we walked to her house; a single small room without access to water and electricity. Her mother, who lived with a new man, had sold the house where Luisa had been living with some relatives. She had hardly anything and while we were heading to her place she was telling me everything she was going to buy in the near future: a table, a cooker and a bed. For some months she had worked as a receptionist at a new company, but in the last two months her salary had not been paid.

Three and half years ago she walked into the Universal Church and stayed. Her mother disliked it very much and their already difficult relationship worsened. Once, Luisa had been ill and her mother consulted a local healer, a *curandeiro*. But Luisa refused the healer's intervention. She recounted that, nevertheless, her mother put the healer's medicine into Luisa's food, which according to Luisa worsened her health. Once she got better she started selling fried chicken and did very well. Her mother became jealous and sabotaged her project. 'But, now, I live on my own and will be able to restart the business', she said and asked for my opinion on her one-woman household, which is perceived as abnormal. When questioned about friends, Luisa stated that she had none and 'in church the women gossip'. The only person she seemed to talk openly to was the pastor. She very much longed to marry. For the last three years she had participated in church services focused on finding a life partner. She had had some boyfriends, but every time she presented them to her mother they never returned.

A month later we met again in church. Luisa had lost weight. She said that she was angry because she had only received a small part of her salary. Her manager held the other part back without justification. She had wanted to buy a table and chairs. 'I am also angry with God', she said, 'I took the envelope for my sacrifice [pastors distribute envelopes to put in one's donations]'. When she kept silent, I questioned her further: 'And you tore the envelope up?' She had not. She said: 'Here I am. Today is the service for prosperity, I cannot fail.'[15] The next time we met she was in a state of shock. The neighbours had shown disrespect to her and she had found various witchcraft medicines before her door. One day,[16] she phoned me. I could not hear her very well. Something was wrong with her. We met in front of the church. She could hardly talk because of throat problems. Luisa explained that her mother was very close with *feiticeiros* (sorcerers) who had given her medicines that burned her throat. She had discovered that her mother wanted to kill her by offering her to a spirit to become rich. Her mother was conspiring with the neighbours, so she had had to move as quickly as possible, but as she was still not receiving her salary the whole situation was complicated. Luisa said that she had prayed a lot with the pastor and made some extra offerings. She found a new temporary home but when, after two months, she could not pay the rent, she was afraid of becoming homeless. Together with the pastor she prepared a *holocausto*. With this special sacrifice she expected to overcome her situation. The pastor's assistants said to Luisa that she should marry: 'you need company'. Luisa said to me, 'but, who can I trust?' She emphasized that she would continue fighting. 'I have hands and feet and will go after a better job and salary.' Since then I have not seen her and have not been able to find her.

I met more women who lived on their own, were desperately looking for a husband and had very tense relations with their kin. It was clear that it was difficult to live alone. Neighbours became suspicious asking why is a woman living alone? Is she involved in evil practices? The women's relative independence complicated their intimate relationships. Suspicions about the intentions from both sides – is he or she after my money? – and about possible spiritual involvements made partners suddenly leave. Luisa, for example, was indeed afraid that she was in the possession of an evil spirit who obstructed her relationships because the spirit 'ate from her wealth' – her sexuality and fertility – as the saying goes. In this context the mother and the spirit were inter-related as they wanted to 'eat' from her. The cases of women like Luisa are extreme examples of the presence of distrust between persons, especially among kin and partners, in establishing control over one's life and over resources. I noticed that, in many families, the members all visited different churches. One went to Catholic mass, another to the Assemblies of God, one to the Zionists and again another to the Universal Church. To be able to prosper, they had to operate individually, otherwise power and money had to be shared. Everyone was looking for a powerful spiritual source for protection against jealousy and harm.

By sacrificing money Luisa on the one hand cut off her relations and dependence on kin, especially her mother. At the same time, she took the lead in her life. She actively participated and offered money during special services focused on finding a partner and a house. For women it was a strategy to control their sexual and marital relationships themselves. Their self-earned money would no longer carry a dangerous potential in their intimate relationships but would contribute to it. Their independence could be converted into 'effective relationships based on love and trust', as propagated by the pastors. Paradoxically, increasing insecurity followed. As the Universal Church strengthened feelings of distrust by emphasizing evil powers in connection to kin, in turn the messages and practices of the church evoked even more reactions of distrust and tension. The church practices trained people to become independent and self-assured in financial and intimate matters; to protect oneself against begging, distrust, jealousy and evil powers. However, this resulted in risk and distrust being fuelled by both insiders and outsiders.

Conclusion: Pentecostalism and development revisited

The apparent emphasis on Pentecostalism as bringing development in the sense of progress and security needs to be critically questioned. Indeed, while the pastors in Mozambique encouraged people to think positively and to start businesses, the techniques they used invariably produced socio-cultural discontinuities, risk and vulnerability. For the upcoming middle classes, to which Pentecostalism was especially alluring, economic liberalism, generating new opportunities as well as new risks, had an enormous effect on reciprocal relationships, building upon former ruptures in the reproductive order that had particularly impacted women. Increased access to waged work and the changes in gender roles shaped a relative independence for the various women who converted to Pentecostalism. Simultaneously, these ruptures brought along uncertainties and vulnerable situations. Jealousy, insecurity and fragile kin and intimate relationships were usually central in members' lives. Contrary to expectations, the Universal Church did not react to this reality by providing a safe shelter, coping strategies, certainty or help to repair relations. Instead, ruptures were increased. Local ideas and practices for gaining prosperity were further broken thereby increasing vulnerability. To escape control by others (kin, partners and spirits) and to achieve control over economic and cultural conditions, an important social order – since the 1990s again supported by Frelimo – had to be destroyed. In this sense Pentecostals obstructed development and progress.

Hence, in this Mozambican context, the consequence of seriously considering that religion is central to the lives of people and is at the heart of social, political and economic processes, underlines the fact that Pentecostalism is intertwined with the neo-liberal order. Pentecostalism does not help people to partake in economic liberalism and to move socially upward – which would posit religion outside social processes (Meyer, 2007) – but is part of its logic. It does not partake in an 'occult economy' (Comaroff and Comaroff, 2000), but erases it and pushes apparent individual choice as the basis of economic liberalism and prosperity. The pastors intensely emphasized and encouraged *personal* development; people should learn to plan their lives, to become independent, to take initiatives and take risks (see also van Dijk, 2009).

However, it would be a mistake to consequently assume that the Universal Church converts were victims of this Pentecostal paradigm. Often the assumption is that churches such as the Universal Church brainwash and exploit people and that the prosperity gospel is wishful thinking

(Gifford, 2004). But converts were not unseeing nor uneducated. They went to the Universal Church exactly because of the pastors' capacities to overrule local powers and dependencies (van de Kamp and van Dijk, forthcoming). It is important to realize that because the church is not offering help, but is part of social transformations it is part of a longer history of breaks. Most converts, for example Dona Silmara, had gone though many changes that went along with a certain upward mobility but had not resulted in a better life. Former breaks called for future breaks. Pentecostalism was a new route to continue the path of transformations that was set in motion a long time ago. The urge to sacrifice so much money, which goes against all kinds of social conventions, developed in the interaction between the pastors and the converts. Interestingly, one pastor told me that the pressure on giving high amounts of money was not like the Church's practice in Brazil. He said that the leaders of the Universal Church were taken by the avidity with which the 'Africans' partook in giving.[17] This was confirmed by older women who had been part of the church since its arrival in Mozambique, like Dona Isabel and Dona Silmara. They said that in recent years the sacrifices had become more extreme. As there seemed to be no limits to the sacrifices converts were prepared to make, the amounts increased. In this sense the Mozambicans and the pastors have reinforced particular aspects of the Pentecostal prosperity gospel.

At the same time, the act of sacrificing not only showed the necessity to erase 'Mozambican culture' once and for all, but it also contained the potential of realizing a new life in the current society. Sacrificing, fighting and struggling were manifestations of individuality, of fully participating in the urban domain and in economic activities. Alternatively, numerous converts dealt pragmatically with their relation to churches. Because the church is not a social institution and trust is not easily cultivated, they left (cf. van Wyk, 2008). Converts hopped from church to church; and this church-hopping may be an example of how converts navigate the field of religion. As the church was part and parcel of a risk society, converts behaved accordingly. Some temporarily participated in the sacrifices to see what they would gain. Converts participated for weeks, months or years in the Universal Church, and then went to a new church. They also learned from earlier experiences. After about 15 years of membership, Dona Silmara left the Universal Church. In the new Pentecostal church she started to frequent, she established a much more independent position towards the pastors. In short, the insecurities of the Universal Church were not an anomaly to development. As part of social, political and economic developments that generate both possibilities and vulnerabilities, Pentecostalism in Mozambique shaped and reinforced them through the interaction of pastors and converts.

Notes

1. Research in Mozambique took place in February 2005, and from August 2005 to August 2007 and August 2008 as part of doctoral research sponsored by the Netherlands Organisation for Scientific Research (NWO). I thank Rijk van Dijk, André Droogers, Anton Houtepen, Miranda Klaver, Kim Knibbe, Regien Smit and the editors for their important comments on earlier drafts of this chapter, which helped to clarify and sharpen the argument. In particular, I am grateful to my Mozambican interlocutors who shared their lives with me. In this text, their names and a few facts about their background have been changed to protect them.
2. Similar practices take place in other Pentecostal churches in Mozambique, like the Brazilian God is Love church and the Portuguese church Maná.
3. For example, several critical articles were published in national journals: 'Multinacional, Comerciante da Fé, Parasite de Deus ou Profeta de Espírito?' (Journal *Savana*, 7 October 1994), 'Acção da IURD em Moçambique' (Journal *Notícias*, 8 April 1997), 'Desactivada Rede Criminosa na Igreja Universal' (Magazine, 9 May 2007).
4. The Universal Church is an urban church and most of its branches are located in Maputo. My research findings in the second city Beira, Central Mozambique, in October 2006, show similarities with the analysis of the Church in Maputo.
5. Interview 15 February 2007.
6. Tithing is donating ten per cent of one's income to the church on a monthly basis. Offerings are extra donations to the church that people are encouraged to give during services as in the above example about the church service focused on success in business. At every service pastors distribute envelopes into which people put their tithes and special offerings.
7. The colour red is a symbol of Jesus' blood that has the power to transform. At the same time this is compatible with the local meaning of red representing a transitory state (cf. Jacobson-Widding, 1989, p. 35).
8. Ethnic group of Southern Mozambique.
9. For similar and related developments in urban South Africa see for example, Ramphele (1993). The urban centres established under European colonialism especially were overwhelmingly male. On the contrary, most West African cities that were much older had more equal numbers of men and women (Sheldon, 1996, p. 6).
10. Conversation 4 February 2007.
11. This percentage is based on my own field data during two years of research. Sometimes 90 per cent or 60 per cent of the church's visitors were women. On average it was about 75 per cent.
12. Conversation 11 January 2006.

13. Interestingly, their daughters are all actively engaged in organizing their *lobolo*, which points to the complex developments of cultural identity in Maputo (see Bagnol, 2006).
14. The pastor knew me as a regular visitor, but did not treat me as a full 'manager'. He nevertheless allowed my presence, because converts took me to the service and the course.
15. 20 November 2006.
16. 5 February 2007.
17. Conversation in May 2007.

References

Arnfred, S. (2001), *Family Forms and Gender Policy in Revolutionary Mozambique (1975–1985)*, Talence: Centre d'Étude d'Afrique Noire.

Asad, T. (2003), *Formations of the Secular: Christianity, Islam, Modernity*, Stanford: Stanford University Press.

Bagnol, B. (2006), *Gender, Self, Multiple Identities, Violence and Magical Interpretations in Lovolo Practices in Southern Mozambique*, Doctoral Thesis, University of Cape Town.

Bähre, E. (2002), 'Witchcraft and the Exchange of Sex, Blood, and Money among Africans in Cape Town, South Africa', *Journal of Religion in Africa*, 32 (3), pp. 300–34.

Belshaw, D., Calderisi, R. & Sugden, C. (eds) (2001), *Faith in Development; Partnership between the World Bank and the Churches of Africa*, Oxford: Regnum/World Bank.

Berger, P. L. (2009), 'Faith and Development', *Society*, 46 (1), pp. 69–75.

Bloch, M. & Parry, J. (eds) (1982), *Death and the Regeneration of Life*, Cambridge: Cambridge University Press.

Brusco, E. (1995), *The Reformation of Machismo: Evangelical Conversion and Gender in Colombia*, Austin: University of Texas Press.

CDE (2008), *Under the Radar: Pentecostalism in South Africa and its Potential Social and Economic Role*, Johannesburg: Centre for Development and Enterprise.

CEA, FM & SARDC-WIDSAA (2000), *Beyond Inequalities: Women in Mozambique, Maputo and Harare*, Centre for African Studies, University Eduardo Mondlane: Forum Mulher and SARDC-WIDSAA.

Comaroff, J. & Comaroff, J. (2000), 'Privatizing the Millennium: New Protestant Ethics and the Spirits of Capitalism in Africa and Elsewhere', *Afrika Spectrum*, 35 (3), pp. 293–312.

Costa, A. B. & Rodrigues, C. U. (2007), 'Famílias e Estratégias de Sobrevivência e Reprodução Social em Luanda e Maputo' in J. Oppenheimer & I. Raposo (eds), *Subúrbios de Luanda e Maputo*, Lisbon: Edições Colibri and Centro de Estudos Sobre África e do Desenvolvimento (CEsA/ISEG/UTL), pp. 139–61.

Cruz e Silva, T. (2003), 'Mozambique' in A. Corten, J.-P. Dozon & A. P. Oro (eds), *Les Nouveaux Conquérants de la Foi. L'Église Universelle du Royaume de Dieu (Brésil)*, Paris: Karthala, pp. 109–17.

de Bruijn, M. & van Dijk, R. (2009), 'Questioning Social Security in the Study of Religion in Africa. The Ambiguous Meaning of the Gift in African

Pentecostalism and Islam' in C. Leutloff-Grandits, A. Peleikis & T. Thelen (eds), *Social Security in Religious Networks. Anthropological Perspectives on New Risks and Ambivalences*, New York: Berghahn Books, pp. 105–27.

de Vries, H. (eds) (2008), *Religion: Beyond a Concept*, New York: Fordham University Press.

Freston, P. (2005), 'The Universal Church of the Kingdom of God: A Brazilian Church Finds Success in Southern Africa', *Journal of Religion in Africa*, 35 (1), pp. 33–65.

Gifford, P. (2004), *Ghana's New Christianity: Pentecostalism in a Globalising Economy*, London: Hurst and Company.

Harries, P. (1994), *Work, Culture, and Identity: Migrant Labourers in Mozambique and South Africa, c. 1860–1910*, Portsmouth, N.H.: Heinemann.

Jacobson-Widding, A. (1989), 'Notions of Heat and Fever among the Manyika of Zimbabwe' in A. Jacobson-Widding & D. Westerlund (eds), *Culture, Experience and Pluralism. Essays on African Ideas of Illness and Healing*, Uppsala: Department of Cultural Anthropology, University of Uppsala and Stockholm: Almqvist & Wiksell International, pp. 27–44.

Lundin, I.B. (2007), *Negotiating Transformation: Urban Livelihoods in Maputo Adapting to Thirty Years of Political and Economic Changes*, Göteborg University: Department of Human and Economic Geography, School of Business, Economics and Law.

Macamo, E. (2005), 'Denying Modernity: The Regulation of Native Labour in Colonial Mozambique and its Postcolonial Aftermath' in E. Macamo (ed.), *Negotiating Modernity: Africa's Ambivalent Experience*, London: Zed Books, Dakar: Codesria Books, Pretoria: University of South Africa Press, pp. 67–97.

Mariano, R. (2003), 'Brésil' in A. Corten, J.-P. Dozon & A. P. Oro (eds), *Les Nouveaux Conquérants de la Foi. L'Église Universelle du Royaume de Dieu (Brésil)*, Paris: Karthala, pp. 45–55.

Marshall, K. & Keough, L. (2004), *Mind, Heart, and Soul in the Fight against Poverty*, Washington, DC: World Bank.

Martin, D. (2002), *Pentecostalism: The World Their Parish*, Oxford: Blackwell.

Mauss, M. (1969) [1924], *The Gift: The Form and Reason for Exchange in Archaic Societies*, London: Routledge and Kegan Paul.

Meyer, B. (1998), ' "Make a Complete Break with the Past": Memory and Postcolonial Modernity in Ghanaian Pentecostal Discourse', *Journal of Religion in Africa*, 28 (3): 316–49.

—— (1999), 'Commodities and the Power of Prayer. Pentecostalist Attitudes Towards Consumption in Contemporary Ghana' in B. Meyer and P. Geschiere (eds), *Globalization and Identity: Dialectics of Flow and Closure*, Oxford: Blackwell, pp. 151–76.

—— (2007), 'Pentecostalism and Neoliberal Capitalism. Faith, Prosperity and Vision in African Pentecostal-Charismatic Churches', *Journal for the Study of Religion*, 20 (2): 5–28.

Newitt, M. (1997), *A History of Mozambique*, London: Hurst and Company.

Oppenheimer, J. and I. Raposo (2002), *A Pobreza em Maputo. A Cooperação Direccionada para os Grupos Vulneráveis no Contexto da Concentração Urbana Acelerada / 1*, Lisboa: Ministério do Trabalho e Solidariedade, Departamento de Cooperação.

Penvenne, J. M. (1995), *African Workers and Colonial Racism: Mozambican Strategies and Struggles in Lourenço Marques, 1877–1962*, Portsmouth, N.H.: Heinemann, London: James Currey and Johannesburg: Witwatersrand University Press.

Ramphele, M. (1993), *A Bed Called Home: Life in the Migrant Hostels of Cape Town*, Edinburgh: Edinburgh University Press.

Ranger, T. O. (ed.) (2008), *Evangelical Christianity and Democracy in Africa*, Oxford: Oxford University Press.

Sheldon, K. (1996), 'Urban African Women: Courtyards, Markets, City Streets' in K. Sheldon (2003), 'Markets and Gardens: Placing Women in the History of Urban Mozambique', *Canadian Journal of African Studies*, 37 (2/3): 358–95.

—— (2002), *Pounders of Grain: A History of Women, Work, and Politics in Mozambique*. Portsmouth, N.H.: Heinemann.

Taylor, C. (2007), *A Secular Age*. Cambridge [etc.]: The Belkamp Press of Harvard University Press.

Van de Kamp, L. and R. Van Dijk (2010), 'Pentecostals Moving South-South: Brazilian and Ghanaian Transnationalism in Southern Africa' in A. Adogame, and J. Spickard (eds) *Religion Crossing Boundaries: Transnational Dynamics in Africa and the New African Diasporic Religions*, Leiden: Brill (forthcoming in 2010).

Van Dijk, R. (1999), 'The Pentecostal Gift: Ghanaian Charismatic Churches and the Moral Innocence of the Global Economy' in R. Fardon, W. Van Binsbergen and R. Van Dijk, *Modernity on a Shoestring: Dimensions of Globalization, Consumption and Development in Africa and Beyond*, Leiden: EIDOS, pp. 71–89.

Van Wyk, I. (2008), 'Anonymity and Public Blessings in a Church of Strangers: The Case of the Universal Church of the Kingdom of God (UCKG) in Durban', Paper presented at the International Conference, *Christianity and Public Culture in Africa*, University of Cambridge, UK, March 2008.

WLSAMOÇ (2001), *Famílias em Contexto de Mudança em Moçambique*, 2nd edn, Maputo: Women and Law in Southern Africa Moçambique.

Part III

'Health Care Provision: Reflections on Religion'

8
Health and the Uses of Religion: Recovering the Political Proper?

James R. Cochrane

Introduction

With the rise of the nation-state and the great corporate ventures of industrial and financial capitalism, especially when considered alongside the decline of religious authority in both of these spheres under conditions of modernity, the role of the citizen has moved to the fore. As the idea of citizenship has expanded, first excluding certain classes, racially defined groups and women, later including them, so too has the question of the role of citizens in society. Specifically, given the expansion of complexity in the spheres of both economy and polity, driven by scientific, technological and managerial innovations, the role of citizens has grown increasingly unstable. It is sometimes attacked or even crushed, sometimes reduced to little more than expressing an opinion now and again through a public vote, and most commonly expressed through the multiple kinds of associational and organizational structures that citizens have themselves built – what we general call civil society (see Cohen & Arato, 1995).

In that context, a key issue has been the extent to which citizens are able to impact directly and with effect upon both polity and economy in large-scale societies under contemporary conditions. Because of what Jürgen Habermas calls the colonization of lifeworlds by the system-imperatives of markets and bureaucracies – by which an instrumental rationality invades and takes over the communicative rationality that is the life-blood of the sphere of the public, reifying anything and everything in the process (including the most intimate spheres of human life) – he fears that an evacuation of the public sphere, or the political proper, is taking place (Habermas, 2001).

This is shifting terrain, as is clear in the differences in the USA between the early years of the recent Bush administration and the movements that helped elect Barack Obama. A great deal seems to depend upon the capacity of ordinary citizens to mobilize and to organize with some freedom and protection. Such conditions are not present everywhere. Meanwhile, one other dynamic, also a mark of the modus operandi of the Bush administration, which may be seen as a peculiarly twentieth-century invention, is that of the skilled and powerfully resourced propagandist (a role first perfected by Goebbels) who is able to utilize the fears and the desires of the populace to great effect in pursuit of an ideological programme. But this has its antecedents, as is perhaps most poignantly captured by Boris Pasternak in *Dr Zhivago*, where towards the end he proclaims that it was the disease, the madness of the age, that everyone was different from his outward appearance and his words.

If trust as credence, and integrity as some basic consistency between word and deed, are necessary for a dependable, durable social contract built through communicative action, then pathological, deliberate and systematic distortions of communication, which have become increasingly acceptable as a pattern of politics in the twentieth century represent a danger. More recently, the danger has grown with the radical turn to an instrumental rationality, under the overarching sign of the 'free' market, which has left nothing sacred, nothing set apart, subjecting everything to the logic of cost-benefit calculations or rational choice theory, thereby undermining the very act of communication itself.

There is little left to relationality here other than exchange relations. The dominant conception of the human being, drawn from the market, is that of an autonomous, self-interested individual who consciously and carefully weighs up costs and benefits in deciding about life. This is a depleted and monological anthropology, and it carries with it at least three potent deficits: that of emptying desire of all but self-gratification or self-fulfilment; that of thinning out what it means to be human to the level of a commodity; and, hence, that of squeezing out from the public sphere the necessary substantive communicative interaction that would secure our life together with justice and dignity. This is the madness of our time.

An alternative paradigm is pursued here, by considering the case of public health in Africa and certain discourses of health that carry a deeper anthropology, via a consideration of the role of religious entities in healthcare. The reasons for focusing on religious entities are also explained. The question posed is whether or not religion might act as one source of vitality for recovering 'the political proper', that discursively redeemed space of the public, currently threatened, in which the communicative interaction necessary to subdue and tame the destructive excesses of political power and the market economy becomes possible.

The health of the public is a particularly compelling lens on society, for health or the lack of it cuts across all human divisions and conflicts, even as it reveals where society itself is diseased (Kim et al., 2000), especially if, following Arjomand (1993, p. 50), the common weal is fittingly understood as *salus rei publicae*, the well-being of the public sphere. Health, as a touchstone *par excellence* of the deficits of system-steering mechanisms in both polity and economy, signals the impact (for better or worse) of system intrusions on the lifeworlds of people (see Hofrichter, 2003). Hence, according to the extent that these lifeworlds are respected and appropriately drawn into any social arrangement or ignored and excluded, it acts as an indicator of good or bad policy. The general relevance of a focus on religion is that lifeworlds in much of sub-Saharan Africa are paradigmatically shaped by religious frameworks.

The argument proceeds by commenting on the history and current state of public health, as a framework for analysing recent research on the contribution of religious health assets to health systems. I shall try to show that paying attention to what is transpiring 'on the ground' in respect of religion and health suggests both a litmus test for the condition of society and a generalizable set of criteria for the reproduction of publicity (which I initially define here as citizens engaged in public discourse). The bias is towards the reconstruction of the political proper 'from below', through transformative agency exercised via communicative action (Habermas, 1984, 1987).

The 'Public' in public health

The determining science of public health – epidemiology – rests upon generalizable, aggregate population-scale measurements of disease or illness, and of the outcomes of specific health interventions. Specific indicators, like infant mortality rates, immunization levels, relative drug costs and the number of hospital beds act as proxies for the health of populations. A powerful and sophisticated range of tools and methods has thus propelled into the foreground the mathematizable constructs of public health as the key drivers of policy and resource allocation.

This has tended to overshadow the original driver of modern public health in the Industrial Revolution: the recognition that disease frequently correlates directly with social and environmental conditions, particularly among the poor. Germ theory played a role, but more important were the interventions in the conditions that encouraged or sustained a disease, such as providing clean water, closing sewage drains or improving working conditions (Rosen, 1999). Religious convictions motivated many people who took up this cause (Gunderson, 1999), prompting, too, the rise of medical missions (Grundmann, 1992, p. 283). More recently, the Christian Medical Commission of the World Council of Churches, with the German Institute for Medical Mission, played a key role in promoting primary health care as a basic policy of the World Health Organisation as proclaimed in its 1978 Alma Ata Declaration (Benn, 2002).

The idea of the common good – a key element of this history – inspired public health precisely at the point where it also began its modern decline. The Industrial Revolution brought with it notions of society that rested on the idea of the sovereign, rationally choosing, autonomous individual. It took the public to be the aggregate of individuals, conceived via definitions of citizenship that were often limited by considerations of merit, race and gender, with the social contract to be guaranteed through generalizable laws which would adjudicate between competing individual interests. The underlying concept of the public was utilitarian. The rise of public health was indeed a direct response to the deficits of a utilitarian view, as Garrett (2000b, p. 585) notes: 'At the dawn of the twentieth century the Western world fused the ideas of civic duty and public health. Conquering disease was viewed as a collective enterprise for the common good.' But that idea faded over time, virtually to disappear towards the end of the twentieth century.

Public health at the end of the twentieth century

Thus, despite inventing many new ways to think about the social dimensions of health (see, for example, Thomas, 2003; Krieger, 2007), public health now faces the prospect of failure for its grand vision of healthy societies (Kim et al., 2000; Scambler, 2002; Farmer, 2003; Hofrichter, 2003; Sanders et al., 2005; Krieger, 2007). Laurie Garrett (2000b) speaks of 'the collapse of global public health'. With rapidly aging populations, increasing environmental loads, resistance to basic and inexpensive drugs, increased vulnerability, larger absolute numbers of people who are injured through accidents or violence and rising problems of mental health, local and regional reversals in public health represent a perplexing setback to the notion of a progressively realized, just and equitable framework for the health of all – at a time when medical science is making discoveries that should make all the difference but do not.

These reversals have something to do with the dominant political economy and its privileging of some over others (see, for example, Garrett, 2000b, pp. 561, 574), producing what health advocate and clinician Paul Farmer (2003) calls 'pathologies of power'. As Amartya Sen puts it, 'The asymmetry of power can indeed generate a kind of quiet brutality' (in Farmer, 2003, p. xvi). Unsurprisingly, the picture is bleakest in sub-Saharan Africa, where the deficits in financial resources, managerial and professional capacities, locally aware and available scientific personnel, facilities, infrastructure and governance plague the attempt to deal with illnesses that are preventable and controllable; where formal health systems are collapsing or in trouble; and where resources are often misapplied or reappropriated.

Besides, the reach of the state is often rather limited, and perhaps decreasing (Holton, 1998; Findlay, 1999; Grugel & Wil, 1999; Habermas, 2001). Compounding these problems is the instrumental rationality by which 'patients become customers, symptoms rather than people are treated, there is an emphasis on productivity rather than care, the system becomes increasingly adversarial ... and the practice of medicine is seen as a *battle* against disease ... rather than the care of the sick person' (Forrester, 1997, p. 162). While the nation-state has not lost its significance (Harvey, 2000, p. 65), transnational realities increasingly define the terrain within which nation-states must act, notably in communication, migration, cultural production, organization, labour and – perhaps above all – finance (Garrett, 2000a).

These dynamics affect the changing fortunes of national public health systems. If the nation-state offers no great hope here, neither do markets. Markets, Habermas notes, though able 'to link efficient and cost-effective information transfer with incentives for expedient information-processing', are 'deaf to information that is not expressed in the language of price'. They also fail to meet the assumptions of perfect competition and equal access upon which the promise of market justice depends: 'Real markets reproduce, and exacerbate, existing relative advantages of businesses, households, and individuals *ex ante*' (Habermas, 2001, p. 95). The inequalities that result, Farmer (1999) argues, are a key indicator in any analysis of the health of the public (see also Scambler, 2002, pp. 86ff.). One could, in fact, say that the essentials of public health constitute human rights (Garrett, 2000b, p. 564). In turn, this suggests that 'its central role is to manage the processes through which the meaning of the health system to society, and so its contribution to broader societal value, is established' (Gilson, 2003, p. 1464).

At stake is a recovery of a politics that discards the neo-liberal doctrine that there is no such thing as society, and reinvents the classical 'republican' vision of the common good in relation to a democratic conception of society. This will mean paying much greater attention to the sources, conditions and potentialities of publicity – of the ability of citizens to enter, with effect, into the public sphere. In health as elsewhere, this implies mediating

> between the necessary leadership or polity from 'above' (*techné*) and the experience and wisdom (*métis*) of those who are 'below', taking into account the asymmetries of power that this equation represents.
>
> (Cochrane, 2009)

Is this possible? To probe this question, I turn to work in sub-Saharan Africa on the nature, role and potential contribution of religious health assets to the health of the public and to strengthening health systems, asking if and how one might understand religion, rooted in the lifeworld, as engendering practices of genuine publicity.

Religion and the political proper

Adequately to relate religion to publicity requires us to clarify the relevant sense of the term religion. Like Durkheim, Habermas (1998, p. 360) sees religion in terms of social integration or cultural reproduction – a functionally separated sphere of identity and belonging in the lifeworld – limiting its significance for the public sphere. To understand religious entities in healthcare, however, requires a perspective interested in the forms of life that represent religion on the ground and the ideas, beliefs or internal logics that drive their action.

This becomes possible with Habermas's recognition that communicative processes aimed at the coordination of action invoke normative self-understandings, including religious ones (see, for example, Habermas, 2002, pp. 148–9). For while religion may aid social integration and the conservation of an established order (Durkheimian), it also evidences internally rooted, historically efficacious and publicly significant embodiments of a counter-tendency, expressed as a transformative potential deeply rooted in the lifeworld.

Still, normative religious self-understandings do construct boundaries and borders of belonging shaped by authority, tradition and institutionalized practices. The social value of belonging, however, also introduces dynamics of inclusion and exclusion. If borders and boundaries are prioritized, then communicative action of the kind required in the public sphere, under modern conditions that enhance publicity, would not be a particular strength of religious communities, unless their self-definition includes an open, porous and flexible sense of borders and boundaries – and only then. This is the crucial clue to identifying which religious persons or groups are most likely to enhance publicity.

Here we might usefully appeal to Derrida's (1992, 2000, 2001) notion of hospitality, the generous reception of the other into one's own bounded space, as one ground for a political ethic aimed at mutuality and justice. How hospitality might open up space for the enhancement of the political proper could be clarified by paying attention to Habermas's suggestions about 'the interaction between two modes for the coordination of social action: "networks" and "lifeworlds" ' It is not the relative hardness or permeability of their boundaries that tells us about how open or closed to participation in the public sphere religious entities might be, but rather, the question of whether the character of their participation reflects a primarily *functional* integration of social relations via networks, or a primarily *social* integration of collective lifeworlds. The former is purposive-rational, aimed at self-interest; the latter is communicative-rational, 'based on mutual understanding, intersubjectively shared norms, and collective values' (Habermas, 2001, p. 82). It is the latter that will enhance publicity, that is, enlarge the sphere of the public and, hence, strengthen the political proper.

Religious health assets: pointers to political potential

We now turn to religion in relation to healthcare, via the notion of religious health assets, with a view to exploring the potential of religion to contribute to the political proper. Is religion of any special relevance as a potential source of publicity? Does religion not militate against plurality and public rationality through a prior commitment to archaic norms and values that foster prejudice or superstition, and an inclination to protect its norms and values from critique or compromise? Do dogmatic or authoritarian traditions not act as a counter to democracy? How, then, can religion contribute to the recovery of the political proper?

Despite its pathologies, religion is also generative in multiple ways, in part the reason for its force, persistence and scale. Does one excise everything religious because of the pathologies, or is it worth paying attention to religion because of its generative capacities and strengths?

The idea of religious health assets places the latter in the foreground, with a pragmatic interest in what one can work with and build upon. Given the scope and scale of religious groups or bodies on the ground, should even 10 per cent of them turn out to be potent sources of publicity, that would make for a powerful movement in most societies. In the context of healthcare, this can be seen in data that comes from research in Lesotho and Zambia by the African Religious Health Assets Programme (ARHAP, 2006) which sought to identify, map and assess religious health assets (RHAs) in the fight against HIV/AIDS (see also http://www.arhap.uct.ac.za). The assets include tangibles (health facilities, personnel, material support, for example) and intangibles (such as volitional, motivational and mobilizing capacities) that carry value for health, and that can be leveraged for greater value. Potentiality (latent possibilities for action) and agency (actualization of such possibilities) are key here.

Using GIS mapping and participatory approaches (Plescia et al., 2001; Craig et al., 2002; Dunn, 2007), and comparing the results with the World Health Organisation (WHO) Healthmapper database, what emerged regarding religious health assets was a picture of a network or ecosystem of religious resources for dealing with HIV and AIDS that substantially exceeded any official (formally actionable) knowledge of interventions. For example, in Chipata in north-eastern Zambia near Malawi, Healthmapper data identified a health centre and a sentinel health site dealing with HIV and AIDS. ARHAP identified an additional 35 entities, all linked to religious bodies: six health centres, 11 congregations (Christian and Muslim) with HIV programmes, five home-based care groups, and three other health support organizations, as well as ten other entities. Overall, of 434 places mapped by ARHAP in Lesotho and Zambia, 432 were officially invisible to the national health system, and almost all of those have a religious character, or partner with religious bodies (African Religious Health Assets Programme, 2006, p. 47).

Also relevant is the scope of their action.[1] Again considering Chipata, this included HIV education, voluntary counselling and testing, prevention of mother-to-child transmission activity, behavioural-change interventions, public advocacy, one case of lab monitoring, one group dealing with compliance strategies, much home-based care, palliative care, support for orphaned and vulnerable children, emotional and counselling support for people living with HIV, treatment support, material support (food, transport and clothing usually) and spiritual support (including prayer and encouragement to be resilient). Several religious entities engaged in 'linking' – in coordinating entities, drawing finance into the area or assisting in fundraising actions, and mobilizing local citizens in support of particular initiatives. Much educational, mobilizing and behavioural-change activity around health is also rooted in the religious life of local communities, through sermons, bible classes, liturgical rituals and internal publications of one kind or another. Such practices provide for a range of communicative acts – rhetorical, textual, performative and demonstrative – that carry with them impulses that awaken and, potentially, mobilize, people around their lifeworld interests.

Tangible religious health assets provide an obvious material base to health interventions, but numerous indicators also point to intangible assets that frequently animate this activity; another focus of the research is aimed at understanding dimensions of religion such as compassion, credibility, trust and commitment that might be critical for scaled-up, locally efficacious and durable interventions. Using a set of action research tools designed for this purpose,[2] some issues relevant to publicity did emerge, including a widespread perception among participants that their struggle for health and well-being is directly related, first, to their context of poverty and a struggle for subsistence, secondly, to weak public health capacity and costly services, and thirdly (in Lesotho), to political instability. Thus many participants saw the growing number of religious entities in health as a direct response to failed state capacities (African Religious Health Assets Programme, 2006, pp. 67, 103).

Furthermore, religious entities frequently operate within an important network of relationships that: (1) integrate their work with secular entities and public health facilities at local level; (2) link them to other religious entities outside of the local context; and (3) tie them to significant intermediary groups, agencies or hubs in the field. These network relationships suggest considerable potential to express publicity and to exercise influence in the public sphere, though many participants also felt insufficiently appreciated as valuable partners in advocacy and policy formulation in the public sphere, suggesting that realizing that potential remains a task still to be fulfilled.

Elements of publicity: against an instrumental view of religion

These findings on religious health assets do not yet tell us if they can or do enhance the public sphere, nor whether their activities are primarily purposive-rational (thus having no direct impact on the public sphere), or communicative-rational forms of coordinated social action and interaction. The ties across local, national and international networks of religious entities might primarily meet self-interested ends, with little implication for publicity. Equally, religious entities might be used by others for their own purposes, as when states seek to transfer costs or responsibilities they should be carrying to private bodies. Either way, there is no obvious reason to believe that the public sphere has been expanded or its discourses enhanced.

Yet the surface may hide much. Links that religious entities have with international bodies, rooted in common religious identity and institutional practices, represent not just networks but also carry cultural, linguistic, historical and political markers of difference that require the respective parties to engage with each other communicatively around differing lifeworlds. Forums, platforms and meetings where religious bodies convene governments, development agencies, scholars, funders and representatives of other religious traditions around strategic action and policy processes are far from uncommon in this context. These international networks are also the medium through which local religious entities often channel their aspirations and desires about their own condition, country and context into public discourse and policy frameworks, through feedback loops from international partners, into their own situation.

Such realities motivate a rapidly growing interest in the human capacities, material resources and demographic reach offered by 'faith-based organizations' in health, as in the WHO's commissioning of research into relevant phenomena (African Religious Health Assets Programme, 2006), which it then used to engage the US Congressional Foreign Relations Committee around the President's Emergency Plan for AIDS Relief (PEPFAR). Similarly, Tearfund and UNAIDS (Haddad et al., 2008), the Centers for Disease Control and Prevention (CDC), the National Institutes for Health (NIH), the Global Fund to Fight HIV, TB and Malaria, and the Bill and Melinda Gates Foundation (Schmid et al., 2008), among others, have begun to pay attention.

Naturally, nervousness accompanies this interest, because to engage with religious- or faith-based bodies draws into the public health discourses perspectives, norms and values potentially incompatible with scientific practice, or potentially a source of conflict rather than a help. Nevertheless, a genuine interest in the lifeworld concerns of religious entities introduces a critical element of public reflexivity, for both the secular and the religious agent must argue for their validity claims with an attempt to persuade. Such interaction, at a minimum, puts the conditions of publicity in place, whatever the outcome. The question is how one maximizes that opportunity for the enhancement of publicity, an expanded sphere of public discourse and practice.

The basic question still remains, however: can we show that religious entities themselves contribute to the opinion and will-formation that is necessary for deliberative democracy? Do they, in principle, acknowledge the validity claims of others with openness to being persuaded by good reasons to change their own position? That religious groups or communities include well-defined, guarded self-understandings, traditional practices and authority structures which tend to work against publicity is common cause, and fear of the other is not a virtue that easily generates publicity.

Yet much that happens in and between religious entities and others in particular contexts suggest potentials for publicity that can be leveraged more effectively. So, for example, at an ARHAP participatory workshop of various regional religious representatives from the Zambian Copperbelt – Christian, African traditional, Muslim, Baha'i and Hindu – to consider their communities' involvement in health in relation to the major health challenges everyone knew of, many met for the first time though they shared many commonalities and concerns. Despite tendencies to defend their own history or positions, as the day wore on, a genuine engagement about differences and similarities in religious lifeworld constructs and beliefs emerged, and numerous people began to discuss how they might coordinate their work for mutual benefit and to influence public policy.

There are many similar examples that can be recounted from ARHAP's research over several years (roughly 40 such workshops in ten countries to date), and no doubt from elsewhere too. This is communicative rationality at work in service of publicity, if one follows Habermas's argument that

> Every encounter in which actors do not just observe each other but take a second-person attitude, reciprocally attributing communicative freedom to each other, unfolds in a linguistically constituted public space.
>
> (Habermas, 1998, p. 361, see also Wuthnow, 1996, pp. 80ff.)

If Habermas is right in suggesting that the social and moral power of religion resides in what is otherwise lost, namely, a 'Sensibility for a miscarried life, for social pathologies, for the failure of individual life plans' (Traub, 2006, p. 15, my translation), then a further dimension of publicity relevant to religious entities is that they are often particularly receptive to what is happening to the health of the people they serve including the broader community. Because of their close connection to individual, private life histories, they possess the appropriate antennae to discern systemic deficiencies.

This is relevant to public health where it aims not only at fulfilling the functions needed for government to deliver health services, but also at the establishment of policies and programmes reflective of the expressed will of the population and enacted where it matters most to them. As Habermas, again, puts it, 'The political public sphere can fulfil its function of perceiving and thematizing encompassing social problems only insofar as it develops out of the communication taking place among *those who are potentially affected'* (1998, p. 365, emphasis in the original).

Here religious entities that embody and represent norms and values of importance to members of the community, propagating and defending people's lifeworld concerns, press for the embedding of the political action system in lifeworld contexts (Habermas, 1998, p. 352). This dual interaction – from the side of the political action system, and from the side of the populace – is critical to generating publicity, and what matters most here is 'the *social space* generated in communicative action' (Habermas, 1998, p. 360).

Conditions of publicity: the integrity of the body

Abu Hamid Huhammed al-Ghazzali once wrote:

> Know that you can have three sorts of relations with princes, governors, and oppressors. The first and worst is that you visit them, the second and better is that they visit you, and the third and safest that you stay far from them, so that neither you see them nor they see you.
>
> (al-Ghazzali cited in Cochran, 1990, p. 168)

This wisdom from a twelfth-century Muslim theologian suggests how ordinary citizens should deal with the public sphere – best avoided. This echoes James Scott's (1991) studies on domination and what he calls the arts of resistance. Publicity is not an automatic desire, and a 'voice' or persuasive presence in the public arena is not guaranteed merely by an ability to communicate, hence enjoin, relationality. It depends upon other conditions being in place too.

Paradigmatically, this includes the integrity of the body, a 'basic particular' of persons (Ricoeur, 1992, pp. 30–5). Harvey (2000, p. 130) notes that 'The body is not monadic, nor does it float freely in some ether of culture, discourses, and representations, however important these may be in materializations of the body', for the self is both voice and body, both linguistically and socio-economically constructed, rooted in both history and expectation (Bloch, 1986a, 1986b). The self capable of entering fully into publicity is one whose imaginative, relational, bodily and material capacity to act is enabled in and with the other, while an incapacity to act is the beginning of the reign of suffering (Ricoeur, 1992).

Human capabilities are thus directly relevant to the conditions of publicity, and they, in turn, are affected by such essential things as safe water, decent housing, good education, a basic income or livelihood, security, bodily integrity and healthy bodies (Sen, 1999; Nussbaum, 2000). Capabilities which include 'Being able to use the senses', 'Being able to have attachments', 'Being able to form a conception of the good', 'Being able to live for and to others' and 'Being able to laugh, to play' (for a useful summary, see Nussbaum, 2001, pp. 87–8). Arguably, 'healthy bodies' is *sui generis*, if one conceives of the body as simultaneously corporeal, mental, spiritual and social (Geurts, 2002; Germond and Molapo, 2006; Germond and Cochrane, 2010), and takes health broadly to be equivalent to comprehensive well-being. On this basis, Harvey's (2000, p. 130) contention is sustained, that 'The body can then be viewed as a nexus through which the possibilities for emancipatory politics can be approached.'

Norms of publicity: decency

To pay attention to the integrity of the body as a basic particular of the person, is simultaneously to recognize personhood, as such, as beyond a biological identity. The need to do so is not new in healthcare, despite the powerful mediation of biomedicine and its inherently impersonal sciences. Thus leading bodies in biomedicine and public health seek solutions to the tension around patient-centred care through ideas such as the 'quality of care' (Institute of Medicine, 2001), 'reciprocity' (Denier, 2005), 'decent care' (Karpf et al., 2008) and 'acceptability' (World Health Organisation, http://www.who.int/hiv/universalaccess2010/). Yet persons, per se, still feature poorly in key decisions about health interventions and resource allocation, particularly those who are on the bottom of the health chain, so to speak – the poor and marginalized, whom we know carry the greatest burden of disease and ill-health. With that recognition, we touch on the question of politics, specifically on a politics from below; that is, the capacity of those most affected by poor health status to participate as citizens in the decisions that affect them most.

The notion of 'acceptability' in the WHO-led campaign for universal access to HIV treatment and care goes some way towards addressing this issue on the demand side of health, focusing on the willingness of the intended recipient to take advantage of what is on offer, to meet critical protocols (especially treatment protocols), and to adjust her or his behaviour accordingly. There is little, if any, focus on the agency or capabilities of the recipient, however, including the right or opportunity to engage in the decision-making and distributive processes that shape health policy.

A move towards a conception of 'decent care' offers much more in the way of active agency on the part of health-seekers and, thus, a stronger likelihood of a viable and meaningful politics of health from below. Echoing the new mandate of the International Labour Organisation (ILO) defined by the idea of 'decent work' (ILO, 2000, p. x; ILO – Report of the Director General, 1999, p. 6), 'decent care' is based in large part on six core values: Dignity, agency, interdependence, solidarity, subsidiarity and sustainability (Ferguson, 2006).

'Dignity' and 'agency' refers to the inclusion of intended recipients of health interventions – such as people living with HIV – in designing and managing their own care processes, while underwriting their capacity to construct and manage their own webs of care. 'Interdependence' and 'solidarity' refer to the integration into the holistic continuum of care of their needs and desires, alongside the capabilities and wisdom of friends and family and the wider community's resources and infrastructure. 'Subsidiarity', locating decision-making action at the lowest appropriate level (where the effects of a decision are most likely to be felt and the competence to take a decision is present), implies that the affected individual is always consulted, and his or her norms and values respected. 'Sustainability' is the assumed outcome of such an approach to localized care programmes, for which there is some evidence in ARHAP's work (in ARHAP's work, see Thomas et al., 2006) and elsewhere (Sanders et al., 2005, p. 758).

Decent care provides a general taxonomy whose particular substance will vary culturally and philosophically from context to context. It rests on a normative, ethical foundation ('the golden rule') that calls one to act towards others as one would expect others to act towards oneself, respecting the dignity, autonomy and relational integrity of the self. Widespread in both secular and religious traditions, such a notion of decency not only describes and prescribes the values and principles about how to treat others in accordance with their full humanity, it also establishes the imperative to abolish all those conditions that would damage or degrade the dignity inherent in all human beings.

(Ferguson, 2006, p. 41)

More than a procedural notion of reciprocity, fairness and equality (the understanding of decency defended by John Rawls, 1999), this conception of decency is de-linked from a particular jurisprudential or socio-political tradition and opened up for concrete, contextually determined interpretations that resonate with a wide range of cultural and religious traditions (Margalit, 1996). To be sure, the proposal does not directly address negative or potentially unhealthy norms such as those associated with patriarchy, nationalism or ethnicism, nor does it take into account that the defence of one's dignity might issue in harmful acts against the other. But its advantage over a procedural approach lies in the careful attention paid to the substantive norms and values that mark a particular person or community, with respect for them and an intention to honour them, yet allowing for a reciprocal justice that finds a limit to what one can do.

As described, decent care addresses the conditions of publicity; it does not, however, tell us about the act of entering the public sphere per se. Nevertheless, it does suggest two things of considerable importance for the growth of publicity in any particular context: it focuses attention on the connection, or lack of it, of state or international policy- and decision-makers to local communities and affected persons; and it encourages local communities and affected persons actively to enter into the processes of policy- and decision-making in respect of their well-being. Because its logic is dialogical, any implementation of a decent care model would, de facto, create a sphere of public discourse filled by interested parties, including the state, civil society and ordinary individuals. Unimplemented, its advocacy remains a political act in itself, one that may be carried out by the affected parties themselves, thus engendering a form of politics from below.

The challenge to religious institutions and leaders

The norms and values of religious communities are readily harmonized with such ideas, and some religious leaders or groups do indeed enter the public sphere around such issues. The presence of religion in the public sphere is to be expected, not just as an empirical fact (Casanova, 1994), but as a philosophical assumption, if one takes seriously a communicative rationality, which asserts the necessity for validity claims, their normative foundations and their value constructs to be part of public discourse.

What really matters, however, is the *nature* of that presence. Participation in public discourse is not yet publicity, which requires that one enter into public discourse ready to defend and enhance that discourse itself as necessary to a healthy society, irrespective of one's own normative claims or value judgements. Similarly, any defence of religious freedom or freedom of belief and conscience means accepting that what is allowed for one must be allowed for all, including the other whose norms and values contradict one's own. That is a condition of publicity. In Habermas's (2002, p. 150) terms:

> Every religious doctrine today encounters the pluralism of different forms of religious truth – as well as the skepticism of a secular, scientific mode of knowing that owes its social authority to a confessed fallibility and a learning process based on long-term revision. Religious dogmas and the attitude of the faithful have to harmonise the illocutionary meaning of religious speech – the affirmation of the truth of a religious statement – with both facts.

There are indeed many clear and persuasive examples of religious entities that do engage in public life on the basis of norms and values that are simultaneously (1) *inclusive* (accepting of the difference of the other), (2) *extensive* (actively going beyond their own community to embrace the other) and (3) *transformative* (open to changing one's own position for the sake of the other). They may enhance community capacity as well, by bringing to bear structural networks, associational patterns, and by appropriately channelling individual perceptions, skills and resources (Norton et al., 2002, p. 205; Aronson et al., 2006). Moreover, religious languages of transcendence are sometimes a crucial defining factor in introducing necessary critical dynamics into politics (Forrester, 1989, p. 81).

What matters, then, if one is concerned about influencing the structural formations of the state (polity) and the market (economy) in ways that reduce their deficits (including their erosion of prized lifeworlds), is to leverage those norms and values internal to religious identities that are inclusive, extensive and transformative. This can only effectively be done by those who stand within a particular community and tradition. The critical issue lies in identifying those religious entities and leaders who most clearly evidence the required sensibilities, those contexts where appropriately leveraging the relevant religious health assets has potential, and then working from there. Given the social reach, scope and scale of religious centres of activity in Africa and elsewhere, and considering the high number of people who belong to a religious communities in the continent, mobilizing even 10 per cent of religious resources in this way – a reasonable possibility in many instances – is likely to have a disproportionally positive effect on public health specifically, and for publicity more generally.

A focus on religious health assets thus illuminates four important things regarding religious entities: (1) they are present on a scale that should not be ignored; (2) their comparatively deep and locally rooted connection to lifeworlds provides a vital source of knowledge of ordinary experience and discourse relevant to the effective implementation of policy; (3) they represent a major component of the health system at both the formal (facility) level and the informal level; and (4) a demographically significant number are likely to be engaged in activities, networks and patterns of communication that, directly or indirectly, have relevance to the enhancement of publicity (Ammerman, 1997, 2005; Bane et al., 2000; Cnaan, 2002).

In Cochran's (1990, p. 168) terms, given the widespread presence of religious entities in health care (our test case), 'The challenge of practical politics is to combine the passion of religion with the civil tolerance of democratic pluralism.' This is more than wishful thinking if considers how many religious traditions have at heart some self-understanding of their task as intrinsically bound to health (Lux and Greenaway, 2006, who describe norms of five faith traditions), by this, meaning the comprehensive well-being of persons, families, communities, societies, and the environment that sustains them. One example: the Christian word 'salvation' (Latin *salus*: to heal, make whole), in long-standing traditional interpretations, refers not just to the person, but to society, the earth and its creatures; in turn, this is rooted in a conception of justice as a condition of well-being (Jennings, 2006).

Within the limiting conditions and strategic considerations I have outlined, and recognizing that religious entities rooted in local communities may be both sensitive to system problems and linked translocally to religious bodies that play a public role, they nevertheless face a common problem for organs of civil society. In Habermas's (Habermas, 1998, p. 370) words, 'the signals they send out and the impulses they give are generally too weak to initiate learning processes or redirect decision making in the political system in the short run', simply because they face political systems with large-scale resources in differentiated functional subsystems whose expert-driven semantic frameworks of decision-making in the face of multiple demands and constrained time and budgetary limits crowd out or sideline civil society.

This, of course, assumes a fairly stable and existing civil society. In many parts of Africa, civil society is weak if not directly threatened and excoriated (Makumbe, 1998). Moreover, viable non-governmental organs, religious entities among them, may themselves play a role alongside, sometimes at one with, the state, in undermining civil society by supporting their political patrons against other actors in civil society.

To imagine that religious entities committed to publicity could influence political systems, then, is a tall order. On the other hand, crisis situations arising from serious legitimacy or delivery problems in the system create some opportunities for swinging the balance somewhat in favour of civil society as people who experience significant life-world deficits or economic hardships bite back at political authorities. Arguably, health, and particularly the effects of collapsing health systems in Africa, constitutes such a crisis. The possibilities for a politics from below that rests upon the centrally important desire for health are perhaps as high now as they have ever been since at least the struggles for decolonization in Africa.

The question and the challenge are to shape such a politics so that it works in favour of a recovery of the political proper in the face of the erosion of the public sphere under contemporary conditions, enhancing the public sphere for the sake of all. It would certainly be an irony for religion to play a role in recovering the political proper, given the widely influential secularist vision that has propounded the exclusion of religion from the public space. But it would not be the first time that the ironies of history have confounded the sceptics and the ideologues.

Acnowledgements

I acknowledge the financial support for various research upon which this paper is based of the Vesper Society, California; the World Health Organisation; the National Research Foundation, South Africa; and the University of Cape Town; noting that none bear any responsibility for my claims.

Notes

1. See African Religious Health Assets Programme (2006), Appendix M, which tabulates the range of activities.
2. Participatory Inquiry into Religious Health Assets, Networks and Agency, or PIRHANA; see African Religious Health Assets Programme (2006).

References

African Religious Health Assets Programme (ARHAP) (2006), Appreciating assets: The contribution of religion to Universal Access in Africa, Report for the World Health Organisation, Cape Town: ARHAP.

Ammerman, N. T. (1997), *Congregation and Community*, New Brunswick: Rutgers University Press.

Ammerman, N. T. (2005), *Pillars of Faith: American Congregations and Their Partners*, Berkeley & Los Angeles: University of California Press.

Arjomand, S. A. (1993), 'Religion and the diversity of normative orders', in Arjomand, S. A. (ed.), *The Political Dimensions of Religion*, New York: State University of New York Press.

Aronson, R. E., Lovelace, K., Hatch, J. W. & Whitehead, T. L. (2006), Strengthening communities and the roles of individuals in community life, in Levy, B. S. & Sidel, V. W. (eds), *Social Injustice and Public Health*, Oxford: Oxford University Press.

Bane, M. J., Coffin, B. & Thiemann, R. (eds) (2000), *Who Will Provide? The Changing Role of Religion in American Social Welfare*, Boulder, CO; Oxford: UK, Westview Press.

Bediako, K. (2000), 'Africa and Christianity on the threshhold of the new millennium: The religious dimension' *African Affairs*, 99, pp. 303–23.

Benn, C. (2002), 'The influence of cultural and religious frameworks on the future course of the HIV/AIDS pandemic' *Journal of Theology for Southern Africa*, 113, pp. 3–18.

Bloch, E. (1986a), *Natural Law and Human Dignity*, Cambridge, Massachusetts: M.I.T Press.

Bloch, E. (1986b), *The Principle of Hope*, Cambridge, Massachusetts: M.I.T Press.

Casanova, J. (1994), *Public Religions in the Modern World*, Chicago: University of Chicago Press.

Cnaan, R. (2002), *The Invisible Caring Hand: American Congregations and the Provision of Welfare*, New York: New York University Press.

Cochran, C. E. (1990), *Religion in Public and Private Life*, New York: Routledge.

Cochrane, J. R. (2009), ' "Fire from above, fire from below": Health, justice and the persistence of the sacred' *Theoria*, 116 (August) pp. 67-96.

Cohen, J. L. & Arato, A. (1995), *Civil Society and Political Theory*, Cambridge: MIT Press.

Craig, W. J., Harris, T. M. & Weiner, D. (eds) (2002), *Community Participation in Geographical Information Systems*, London: Taylor and Francis.

Denier, Y. (2005), 'Public health, well-being and reciprocity' *Ethical Perspectives: Journal of the European Ethics Network*, 11, pp. 41–66.

Derrida, J. (1992), *The Other Heading: Reflections on Today's Europe*, Bloomington: Indiana University Press.

Derrida, J. (2000), *Of Hospitality*, R. Bowlby (trans.), Stanford: Stanford University Press.

Derrida, J. (2001), *On Cosmopolitanism and Forgiveness*, New York: Routledge.

Dunn, C. E. (2007), 'Participatory GIS – a people's GIS?' *Progress in Human Geography*, 31, pp. 616–37.

Farmer, P. (1999), *Infections and Inequalities: The Modern Plagues*, Berkeley; Los Angeles: University of California.

Farmer, P. (2003), *Pathologies of Power: Health, Human Rights and the New War on the Poor*, Berkeley: University of California Press.

Ferguson, J. T. (2006), 'Decent care: Towards a new ethic in the provision of care', unpublished, Presentation, World Health Organisation Global Consultation on Decent Care, Vervey: Switzerland, pp. 26–30 (June).

Findlay, M. (1999), *The Globalisation of Crime: Understanding Transitional Relationships in Context*, Cambridge: Cambridge University Press.

Fleßa, S. (2002), *Gesundheitsreformen in Entwicklungsländern: eine kritische Analyse aus Sicht der kirchlichen Entwicklungshilfe*, Frankfurt am Main: Verlag Otto Lembeck.

Forrester, D. B. (1989), *Beliefs, Values and Policies: Conviction Politics in a Secular Age*, Oxford: Clarendon.

Forrester, D. B. (1997), *Christian Justice and Public Policy*, Cambridge: Cambridge University Press.

Garrett, G. (2000a), 'The causes of globalization' *Comparative Political Studies*, 33, pp. 941–91.

Garrett, G. (2000b), *Betrayal of Trust: The Collapse of Global Public Health*, New York, Hyperion.

Germond, P. & Cochrane, J. R. (2010), 'Healthworlds: Conceptualizing landscapes of plural healing systems' *British Journal of Sociology* (forthcoming), originally presented at the conference on Reasons of Faith: Religion in Modern Public Life, University of Witwatersrand: Wits Institute for Social and Economic Research (WISER), 2005.

Germond, P. & Molapo, S. (2006), 'In search of Bophelo in a time of AIDS: Seeking a coherence of economies of health and economies of salvation' *Journal of Theology for Southern Africa*, 126, pp. 27–47.

Geurts, K. L. (2002), *Culture and the Senses: Bodily Ways of Knowing in an African Community*, Berkeley: University of California Press.

Gilson, L. (2003), 'Trust and the development of health care as a social institution' *Social Science & Medicine*, 56, pp. 1453–68.

Grugel, J. & Wil, H. (1999), *Regionalism Across the North-South Divide: State Strategies and Globalization*, New York: Routledge: Methuen.

Grundmann, C. H. (1992), *Gesandt zu heilen! Aufkommen und Entwicklung der ärztlichen Mission im neunzenten Jahrhundert*, Gütersloh: Gütersloher Verlagshaus Gerd Mohn.

Gunderson, G. (1999), 'Good news for the whole community: Reflections on the history of the first century of the social gospel movement', presented as the Earl Lecture, *Pacific School of Religion*, 28 January edition.

Habermas, J. (1984), *The Theory of Communicative Action Volume 1: Reason and the Rationalization of Society*, Cambridge: Polity.

Habermas, J. (1987), *Lifeworld and System: A Critique of Functionalist Reason*, Boston: Beacon.

Habermas, J. (1998), *Between Facts and Norms: Contributions to a Discourse Theory of Law and Democracy*, Cambridge, MA: MIT Press.

Habermas, J. (2001), *The Postnational Constellation: Political Essays*, Cambridge: Polity Press.

Habermas, J. (2002), *Religion and Rationality: Essays on Reason, God and Modernity*, Cambridge: Polity Press.

Haddad, B., Olivier, J. & De Gruchy, S. (2008), The potential and perils of partnership: Christian religious entities and collaborative stakeholders responding to HIV and AIDS in Kenya, Malawi and the DRC, Cape Town & Pietermaritzburg, African Religious Heath Assets Programme, commissioned by Tearfund and UNAIDS.

Harvey, D. (2000), *Spaces of Hope*, Berkeley: University of California Press.

Hofrichter, R. (ed.) (2003), *Health and Social Justice: Politics, Ideology, and Inequity in the Distribution of Disease*, San Francisco: Jossey-Bass [John Wiley & Sons].

Holton, R. J. (1998), *Globalization and the Nation-state*, Houndmills, Basingstoke, Hampshire: Macmillan.

ILO – Report of the Director General (1999), 'Decent work', Geneva: International Labour Organisation.

ILO (2000), 'Decent work and poverty reduction in the global economy', *Preparatory Committee for the Special Session of the General Assembly*, 2nd session edn, New York: International Labour Office.

Institute of Medicine, C. o. Q. o. H. C. i. A. (2001), *Crossing the Quality Chasm: A New Health System for the 21st Century*, Washington, D.C.: National Academy Press.

Jennings Jr., T. W. (2006), *Reading Derrida/Thinking Paul: On Justice*, Stanford, California: Stanford University Press.

Karpf, T., Ferguson, J. T., Swift, R. & Lazarus, J. V. (eds) (2008), *Restoring Hope: Decent Care in the Midst of HIV/AIDS*, London: Palgrave Macmillan.

Kim, J. Y., Millen, J. V., Irwin, A. & Gershman, J. (eds) (2000), *Dying for Growth: Global Inequality and the Health of the Poor*, Monroe, ME: Common Courage Press.

Krieger, N. (2007), 'Why epidemiologists cannot afford to ignore poverty' *Epidemiology*, 18, pp. 658–63.

Lux, S. & Greenaway, K. (2006), 'Scaling up effective partnerships: A guide to working with faith-based organizations in the response to HIV and AIDS', Oxford: Ecumenical Advocacy Alliance.

Makumbe, J. M. (1998), 'Is there a civil society in Africa?' *International Affairs*, 74 (2), pp. 305–19.

Margalit, A. (1996), *The Decent Society*, Cambridge: Harvard University Press.

Norton, B. L., McLeroy, K. R., Burdine, J. N., Michael, R., Felix, J. & Dorsey, A. M. (2002), 'Community capacity: Concept, theories, and methods', in Diclemente, R. J., Crosby, R. A. & Kegler, M. C. (eds) *Emerging Theories in Health Promotion Practice and Research: Strategies for Improving Public Health*, San Francisco: Jossey-Bass [John Wiley & Sons].

Nussbaum, M. C. (2000), *Women and Human Development: The Capabilities Approach*, Cambridge: Cambridge University Press.

Nussbaum, M. C. (2001), 'Adaptive preferences and women's options (Symposium on Amartya Sen's Philosophy: 5) *Economics and Philosophy*, 17, pp. 67–88.

Plescia, M., Koontz, S. & Laurent, S. (2001), 'Community assessment in a vertically integrated health care system' *American Journal of Public Health*, 91, pp. 811–14.

Rawls, John. (1999), *The Law of Peoples*, Cambridge: Harvard University Press.

Ricoeur, P. (1992), *Oneself as Another*, Chicago: University of Chicago Press.

Rosen, G. (1999), *A History of Public Health*, Baltimore: Johns Hopkins University Press.

Sanders, D. M., Todd, C. & Chopra, M. (2005) 'Confronting Africa's health crisis: More of the same will not be enough' *British Medical Journal*, 331, pp. 755–8.

Scambler, G. (2002), *Health and Social Change: A Critical Theory*, Buckingham and Philadelphia: The Open University Press.

Schmid, B., Thomas, E., Olivier, J. & Cochrane, J. R. (2008), 'The contribution of faith based organizations and networks to health in sub-Saharan Africa, Cape Town' African Religious Health Assets Programme, Study commissioned by B & M Gates Foundation.

Scott, J. C. (1991), *Domination and the Arts of Resistance: Hidden Transcripts*, New Haven: Yale University Press.

Sen, A. (1999), *Development as Freedom*, New York: Random House.

Thomas, L., Schmid, B., Cochrane, J. R., Gwele, M. & Ngcubo, R. (2006), ' "Let us embrace": Role and significance of an integrated faith-based initiative for HIV and Aids: Masangane case study, Cape Town' African Religious Health Assets Programme: University of Cape Town.

Thomas, R. K. (2003), *Society and Health: Sociology for Health Professionals*, New York: Kluwer Academic/Plenum Publishers.

Traub, R. (2006), 'Der Einfluss der Religionen auf Politik und Gesellschaft' in *Spiegel Special*, Weltmacht Religion: Wie der Glaube Politik und Gesellschaft beeinflusst, pp. 6–15.

Wuthnow, R. (1996), *Christianity and Civil Society: The Contemporary Debate*, Valley Forge: Trinity Press International.

9
Marshalling the Powers: The Challenge of Everyday Religion for Development

Elizabeth Graveling

Introduction: religion and development

Over the past decade development theoreticians and practitioners have shown increasing interest in religion. In the same year as Ver Beek (2000) labelled religion 'a development taboo' the World Bank organized, with the Council of Anglican Provinces of Africa, a conference in Nairobi entitled *Faith in Development: Partnership between the World Bank and the Churches of Africa* (documented in Belshaw et al., 2001). Four years later the UK Department for International Development announced a five-year research programme on 'Faiths in Development' (which became the 'Religions and Development' group based at the University of Birmingham). This recent attention has reflected a wider renewed interest in religion across the social sciences (see for example Petito and Hatzopoulos's (2003) *Religion in International Relations: The Return from Exile*). This has been attributed to a number of reasons, including the emergence since the 1960s of the 'developing world' with diverse cultural and religious heritages in a context of dramatically increasing global communications (Haynes, 1993), and the recent rising awareness of Islam and religious fundamentalism. Thomas interprets the global resurgence of religion as 'part of the larger crisis of modernity' (2003, p. 22). It reflects a disillusionment with the 'modern' ethos which refuses to recognize the existence or relevance of anything outside the positivist epistemology of science and rationality, as well as a failure of modernization to produce democracy and development in less-developed countries and a 'struggle for cultural liberation, or the global struggle for authenticity', as part of a 'revolt against the West' (ibid., p. 22). It is also linked with contestations of the modernist secularization thesis, indicating that religion in Western societies may be transforming rather than declining (Cameron, 2003; Heelas & Woodhead, 2005), and that elsewhere it is expanding rapidly (Barrett, 2001; Martin, 2001). There is increasing awareness that 'modernity' is not a universal state to be arrived at through a standardizing process of modernization, but that there are 'many modernities' (Comaroff & Comaroff, 1993, p. ix), to many of which religiosity and spirituality are intrinsic.

Religion has been involved and implicated in development from the first 'civilizing' missionary endeavours in pre-colonial times which included attempts to provide education and healthcare – well before 'development' emerged as a discipline and an industry. However, mainstream development has roots in a modernist agenda based on post-Enlightenment principles. The secular and the sacred are separated and opposed along the same lines as physical and spiritual, rational and irrational, modern and traditional, public and private. It is only with the recent shift towards postmodernism and the opening up of development as a more contested and pluralized arena that mainstream secular development practitioners and theoreticians have begun to view religion as playing a meaningful role both in development itself and in the lives of people targeted by development activities. Over the past decade there has been growing interest in engaging with religion, both as a means to achieving established development goals and as it is increasingly recognized that religion is central to the lives and values of most people in developing countries, thereby also playing a potential role in shaping development programmes (Belshaw et al., 2001; ter Haar & Ellis, 2006; Tyndale, 2006; Marshall & Van Saanen, 2007; Rakodi, 2007). This engagement has, however, largely been sought on an organizational level, with development agencies urged to interact with specific 'faith communities' (Belshaw et al., 2001) and 'faith-based organizations' (FBOs) (Clarke, 2006). Of course, the organizational aspect of religion is important: this is where rulings on doctrine and policy are made, where the voice of members may be expressed, and where the potential for dialogue and co-operation with external agencies is found. However, such organizations and 'communities' do not necessarily represent discrete bodies of people with coherent beliefs and secure boundaries. There may be enormous differences between the official policy of a church as set out by its head office and the activities of local branches; there may also be great divergence between the doctrines and teachings of a local church and the beliefs, values and practices of its members. Attending or belonging to a church does not necessarily entail full acceptance of or compliance with its values and teachings; nor does it exclude acceptance of or engagement with other religious or non-religious traditions. Members of churches may interpret and draw on religious discourses in many different ways, influenced by other socio-cultural discourses within which they live. Moreover, the composition of congregations and 'faith communities' is neither uniform nor constant: firstly, members of the same church may differ vastly in their relationship with that church; secondly, members may move between – or in and out of – groups, drawing on discourses from different churches and even different religions either sequentially or simultaneously.

In order for the development world to engage fully with religion, then, it must not only consider possibilities for development contributions and collaborations among religious organizations, but must also be meaningful and relevant to the religious reality of people's daily lives. This calls for, as well as analysis focussing on FBOs and religious organizations, research that considers the complexities and ambiguities of religion in everyday life. This chapter draws on such research, carried out in the context of Ndwumizili, a village in southern Ghana, which explores how religion is experienced and constructed in the lives of members of two of its churches.[1] I first give an overview of academic approaches to Christianity in Ghana and then discuss the worldview of the residents of Ndwumizili and the implications of this for their strategies in addressing problems, drawing on literature from witchcraft studies and medical anthropology.

Approaches to Christianity in Ghana

The tendency to work in terms of discretely bounded religious groups is not unique to development. Deciding where to lay lines of division and categorization has been a major issue in the study of Christianity in Ghana, and in Africa more generally. There are an enormous variety of churches and religious movements which have proliferated over recent years, inviting attempts to categorize and fit them into typologies (Fernandez, 1978, pp. 201–6; Foli, 2001; Gifford, 2004). To the broadly recognized categories of mission/historic churches, African Indigenous Churches (AICs) and (classical) Pentecostal churches have been added the neo-Pentecostal or charismatic churches which have burst onto the scene over the past 25 years. The majority of work on Christianity in Ghana has been based on these four areas, usually concentrating on just one. Despite acknowledgement by scholars of the problems inherent in discussing 'types' of churches such as mission churches, AICs or Pentecostals, because of both the blurred boundaries between them and the variety within them (e.g. Gifford, 2004; Meyer, 2004), the tendency – even by the same authors – is to continue to do so, thus upholding the reification of categories some of them seek to deconstruct (Meyer, 2004). Discussion thus remains focussed around denominational discourses, which does not allow for the recognition of fluidity between different church and non-church groups, nor of the creative and often eclectic ways in which people interpret and negotiate these discourses.

It is not only between church denominations that artificial oppositions have been imposed. The study of Christianity in Africa has been underpinned by the Eurocentric view of Christianity as a Western religion introduced by missionaries into a foreign (African) culture. Studies of African indigenous churches in particular have therefore centred on the Africanization of Christianity and the Christianization of African culture (see e.g. Fasholé-Luke et al., 1978), an approach which, as Meyer (2004) argues, reifies Africa and Christianity as oppositional categories which need to be reconciled. Pentecostal-charismatic churches, on the other hand, are approached in the context of discussions regarding the relationship between two other categories, modernity and tradition, whether seen as trapping people in a worldview based on supernatural rather than technical-rational principles (Gifford, 2004), or as space providing the potential to break with the past and engage with the modern (Meyer, 1998).

We are therefore faced with a series of binary oppositions, such as secular-sacred, physical-spiritual, modern-traditional, Christianity-culture and Western-African. On top of these are sets of categories, primarily different church sectoral and denominational groupings and different religions. Even when these are not viewed as static and it is recognized that the way they are conceptualized, practised, articulated and constructed varies according to spatial, temporal, environmental and social conditions, it is still the discourse or religion that frames analysis. In everyday life, however, people do not always think in terms of discourses. Daily life tends to be focussed around daily, existential concerns, as has been repeatedly shown in the case of Ghana (Larbi, 2001; Sackey, 2001; Gifford, 2004): money, health, education, status, marriage and children are pressing issues.

Moreover, people do not necessarily refer to a fully worked-out, coherent and identifiable belief system or cosmology in making decisions regarding their actions. Culture is not static and discrete, but rather fluid and dynamic, with customs and traditions being continually adapted according to new circumstances and ethnic groups intermingling and sharing cultural values and practices. It is therefore impossible to talk of an 'authentic' African, Western, or even British, Yoruba or Ahanta culture.

Ndwumizili and its churches

The village of Ndwumizili lies in the Ahanta West district of Ghana's Western Region, an hour's journey from Sekondi-Takoradi, the regional capital. Ndwumizili has a population of approximately 2000, and its main industries are agriculture, fishing and tourism, visitors being attracted to its long, sandy beach. The village is composed mainly of a tangle of mud and thatch buildings alongside concrete structures with corrugated iron roofs. Although there is an electricity supply (subject to frequent power cuts) to which many houses are connected, there is no pipe-borne water and most of the village relies on a few public and private wells. The village has no health facilities beyond a basic pharmacy run by a former nurse. The school, which caters for about 600 children at kindergarten, primary and junior secondary levels, has only basic equipment.

The predominant ethnic group in Ndwumizili is Ahanta, a minority ethnicity related to the majority Akan people who cover most of the south of Ghana. However, members of other groups have also settled in the area, predominantly Fantes and Nzemas who occupy the neighbouring land to the east and west of Ahantaland respectively. Ndwumizili is governed by a combination of the traditional Ahanta chieftaincy, inherited through a matrilineal system, and the Ghanaian state government, of which the elected assemblyman is the local representative. Apart from these institutions of authority, the village possesses few organized bodies of people. The most striking, due to their sheer number, are the churches, of which there are 11 in total. Each represents a different denomination, falling into three broadly recognized classes (see for example Foli, 2001): historic or mission churches (Methodist, Roman Catholic and Seventh Day Adventist); Pentecostal churches (Church of Pentecost, Deeper Life Bible Church, Assemblies of God, Christ Apostolic Church, Christ Bethel Mission International and the Action Church); and 'spiritual' churches or AICs (Musama Disco Christo Church and Church of the 12 Apostles). The study on which this paper is based focuses on two of these: the Musama Disco Christo Church (MDCC) and the Assemblies of God Church (AG).

Churches both reflect and influence people's worldviews. The two at the centre of this study differ greatly in their history, organization and practices, to the extent that one (the AG) does not consider the other properly Christian. The MDCC is an indigenous Ghanaian church. Although it now has branches in other countries, including the United Kingdom and the United States of America, its organization closely mirrors the social structures of the culture in which it was founded. It also imposes several markers of membership, including wearing particular jewellery, paying a monthly fee and learning a special language. MDCC worship revolves around fixed liturgy learnt by heart and corporate rituals including much singing, dancing and drumming as well as the symbolic use of items such as water, candles, incense, oil and crosses. The church leadership is strictly hierarchical, knowledge is esoteric and pastors and prophets revered. This is in contrast to the AG, which was originally an American mission church and has a more democratic structure based on notions of equality, though pastors are still highly respected and deferred to as 'men of God'. This church is based on regular individual and corporate study of the Bible. It opposes ritual and the symbolic use of items such as those mentioned above. Rather than rooting itself in local culture, it seeks to transcend culture and sees itself as part of a global community of not only AG but all Pentecostal churches.

A world of powers

Rather than conceptualizing the world in terms of separate religious discourses, each with its own cosmology and spiritual actors, residents of Ndwumizili appear to view the universe as a continuous landscape populated by a range of different powers, both spirits and human. These include a supreme creator God, secondary deities or 'small gods', ancestors, evil spirits, the devil, witches and juju practitioners. They also include doctors, herbalists and institutions of authority such as the chieftaincy, the state, the family and the church. Few (if any) people in Ndwumizili, whatever their 'religion', would deny the existence of any of these powers, or that each of them has the capacity to intervene in or influence one's life, and most of them are understood as being able to bring both blessings and harm. Although they are not all equally powerful or united in purpose (indeed, they may be as opposed to each other as good and evil), they do not 'belong' to separate religions and belief in that their existence does not depend on acceptance of a particular religious tradition. There is no question but that they all exist: the question is how and to what extent to interact with them. Each person must therefore negotiate this world of powers, placing him or herself in the best possible position in relation to each of them in order to gain maximum benefit, protecting themselves from harm and procuring blessings.

The way these powers are viewed is not fixed or uniform. There have been shifts over time as new powers have entered the arena, such as biomedical doctors, the nation-state and Jesus, while others have gained new prominence, such as the perceived presence of witches. Related to these and wider structural socio-economic changes, the perception of established powers has altered, particularly the 'small gods' and to a lesser extent the ancestors, both of whom appear to have moved from a morally neutral position, concerned with both their own interests and the upkeep of morality and social cohesion, to being viewed by many as inherently evil. This reconceptualization has been brought about largely by the growth of Christianity, particularly Pentecostalism which, rather than denying the existence of such powers, designates them as evil, aligned with the devil and opposed to God (Hackett, 2003; Ellis & ter Haar, 2004). However, as change is always ongoing, such reframing is never complete. In people's perceptions of and dealings with powers – not only 'small gods' and ancestors – there is a sense of uncertainty. This is not only because many of the powers are largely invisible and incorporeal. Their nature is contested and usually ambivalent; they are seen differently from different perspectives and through time, and they may bring both harm and good. They also tend to be personified and personalized; negotiating the powers entails engaging in relationships with beings that have their own interests and agenda – and relationships are never entirely predictable. The powers exist as influences and options and people act pragmatically in exercising these options, but the outcome is never certain.

Marshalling the powers

Money, church and witchcraft

One type of power which plays an important and increasing role in everyday life in Ndwumizili is witchcraft. Witchcraft has long been dis-associated with tradition and 'primitive' thinking and associated with social upheaval and modernity; indeed, Moore and Sanders observe that 'Contemporary scholars of witchcraft cast occult beliefs and prac-tices as not only contiguous with, but constitutive of modernity' (2001, pp. 11–12). The reason for the recent heightened interest in witchcraft and modernity is that reports of witchcraft have been increasing over the past few decades, rather than decreasing as modernists might have expected (Ciekawy & Geschiere, 1998). One of the characteristics of modernity is the opening up and meeting of new and different markets, political systems, cultures and technologies; people are required to negotiate undefined foreign powers of global capitalism, international and state politics, hi-tech communications and transport. There are new opportunities for the creation of wealth and new frustrations for those who cannot take advantage of them. As Comaroff & Comaroff assert

> the signs and practices of witchcraft are integral to the experience of the contemporary world. They are called on to counter the magic of modernity. And to act upon the elusive effects of transnational forces – especially as they come to be embodied in the all-too-physical forms of their local beneficiaries.
>
> (1993, pp. xxv)

In the face of a pluralistic, non-coherent and rapidly changing society, where people live in a nexus of many different – often discordant – ideas, values and practices, witches, who act and interact with unseen powers, constitute a very plausible interpretive framework. In Nigeria, Bastina (1993) highlights the way that tensions based on the exploitation and expropriation of resources both between indigenous residents and non-indigenous residents, and between resident indigenes and non-resident indigenes of villages are often constructed in terms of witchcraft. From the rural perspective, city-dwellers resemble witches in their separation from the village community, their unregulated accumulation of wealth and their association with cities which continually draw resources from rural areas. From the urban perspective, on the other hand, the village is the abode of witchcraft and rural residents act as such when they accept money from urban relations while refusing them full status in the community. Modernity is thus negotiated partly through spiritual discourses.

In Ghana there is an ongoing media debate over the place of money in church. There are regular reports of scandals involving 'fake' pastors who defraud innocent members of the public out of enormous sums of money by charging them for prayer and miracles.[2] This can be related to the pervasive 'prosperity gospel' taught with sincerity in many churches, which asserts that health and wealth are directly proportionate to faith, and that the more one gives to God, the more one will receive. This often results in extremely rich church leaders, inciting resentment and cynicism from the media and the public. It is not just independent pastors that come in for criticism; leaders of established denominations such as the Catholic and Methodist churches are also criticized for their relative wealth: 'Every week you have to give money, offering, offering, and then you go build a big house, and a poor man like me, I can't build a house but I have to give' (Kwame, ex-MDCC but speaking of churches in general). However, although the pastors of the MDCC and AG churches in Ndwumizili both receive their income directly from church offerings, neither is wealthy. The AG pastor relies on his wife's teacher's salary and living accommodation, and the MDCC pastor lives off earnings from her guest house and pharmacy. Both appear to contribute more financially to their respective churches than they receive: the MDCC pastor has funded the majority of the ongoing church building project from her private income, while the AG pastor and his wife have paid for all their church's chairs. Both pastors also receive frequent requests for financial and material assistance from needy members of their congregations and sometimes from others outside the church. It is not surprising that, while leaders of larger churches in urban centres can benefit enormously from such a system of remuneration, rural pastors whose congregations are mostly very small and very poor struggle to find the means to make ends meet.

It is clear that money plays a large role in churches. At a typical service in either the MDCC or the AG there are at least two offerings, plus announcements, exhortations and discussions on issues of church finances and contributions from members. In practice, of course, it is not easy to raise large amounts of money from poor congregations. A multiplicity of demands results in less being contributed per offering or, where sums are fixed, payments are deferred and people participate, in effect, on credit. During MDCC services the pastor occasionally reads out names of those who have failed to pay their dues for several months. However, despite regular warnings, it is rare for members to be expelled from church for defaulting on financial contributions. More common in both churches is self-exclusion, where people do not attend church because they cannot contribute to offerings. Excessive financial demand is the most common reason given for leaving or not attending church, covering virtually all denominations. For some this is linked to a sense of disillusionment and cynicism towards the church, particularly where pastors are perceived as accumulating wealth from the efforts of their congregations. It is also commonly related to concern for one's image in society and fear of being judged and gossiped about. Most offerings are carried out in a very public manner so that it is easy to feel self-conscious about the amount one gives. In a context where both rich and poor are potential targets of suspicions and accusations of witchcraft (the former for using witchcraft to feed their avarice and the latter for harming or preventing others from prospering, through jealousy), managing one's reputation is important. Relationships tend to be ambivalent: it is assumed that one has enemies, but these are not always easily identifiable and may be found among those one is close to, including friends, family and church members. People are therefore wary of revealing their hopes, plans and financial affairs and gossip is greatly feared, due to possibilities both of attracting accusations of witchcraft and of exposing oneself to harm by witches.

A wealthy church attracts as well as repels, offering useful social connections and promises of riches. Such a church can be understood not just as attended by wealthy people, but as a principal factor in their success, therefore attracting other people who seek to receive similar blessings. Prosperity implies power: the wealthier the pastor, the greater access to power they are perceived to have – although precisely which power they obtain their wealth through is not always clear. Status as a church leader does not protect one from suspicion of engaging in witchcraft or juju, but evidence of any kind of spiritual power is viewed with caution and respect.

Moreover, in both churches, giving money is presented as part of a solution to problems – it is through giving to God that one is blessed. Leaders and members of both churches continually speak in terms of blessings and protection from harm. Blessings come from God and harm from the devil, and the absence of one is often interpreted as the presence of the other. Blessings and protection can be attained through obedience to God and living holy lives (including giving to God through the church), through constant prayer and, in the MDCC (and particularly for protection), through participation in church rituals. Therefore, although attending church may add to people's problems it is also seen as a potential way out of them. If monetary demands are the biggest cause of non-attendance, the most commonly cited reason for attending church (after family influence) is to find a solution to one's problems. These problems very commonly relate to poverty, thus their solutions are often financial. The relationship between church and problems is therefore ambivalent, the former being both a contributory factor to the latter (through giving), and a potential solution (through receiving) – and giving is often presented as the prerequisite for receiving.

Fluidity between churches: Vincent, Abena and Beth

Problems come in many shapes and sizes: people attend church in search of solutions to domestic conflict, poverty, ill-health, exams, infertility, singleness and unemployment among other things. Church is not the only source of help and as people seek specific blessings and protection from harm they may explore and combine different strategies. They may also explore different churches, which compete in the market of providing answers to problems. Very few people attend only one denomination of church throughout their life; most switch at least three times. For some, their participation in a church depends directly on the extent to which it can meet their existential needs. Vincent, aged 20, identifies the AG as his preferred church although he currently does not attend meetings. He has attended several different denominations; throughout his childhood this depended on the family member he was living with at the time (with his aunt he attended an Anglican church, with his mother an MDCC and then a Roman Catholic church, and with his father a Methodist church), but more recently he has moved into the Pentecostal sector. Here he describes his reasons for attending Shiloh and Liberty, two Pentecostal churches in Takoradi:

> I went to this church because my two brothers were also going to a church. Before they were going to Methodist, then they stopped and go to Shiloh, this church. So . . . they came to me and said Vincent, why don't you come to our church, because the pastor is a good pastor, if you have any difficulties the pastor can help you. So I say okay.

> I went to Liberty because at that time I was having a problem . . . . And this pastor was preaching at a radio station. So I heard his voice on the radio and I think he can solve the problem for me. So I wake up in the morning, it was Wednesday, then I went to this church. When I went there they welcomed me and those ushers, those people at the church said if you are having any problem you should come. So I say okay I have some problem to discuss with them and they can help. So they called me and I just took a paper like this and they write everything that I was saying, they were writing it down. And they give it to the pastor and they give me a seat to sit on. I was going to this church for prayers. [. . .] I give a testimony that God has listened to my prayers, because I was in difficulty, and now I get someone who has helped me to get what I need. But the other problem was not solved. I went there with two problems . . . one for my father, I want my father to come [from Nigeria], you know? And one for a friend from America who wants to help me to get a visa to there . . . . And these problems, one worked but the other one did not work . . . . My

father did not come. But my friend from America, I heard from him, it was a long time, I had written a lot of letters, I hadn't heard, you know? So I heard from him ... Yeah, I heard from him, everything was working then, they were making all the papers, you know? And later also I did not hear from him again. From there I stopped [attending] Liberty.

(Vincent, 4 May 2006)

Vincent also explains his reasons for liking the AG:

They were preaching the goodness of God, you know? And they were also teaching something from the Bible that says we should love our neighbour as ourselves, we shouldn't steal, we should be a good man of our faith, you know? So this actually helped me a lot.

They give me a special place to sit, you know? Though I was young, but they encouraged me, they told me oh, you can come to Sunday school. So I get to know the people want me to be somebody in church.... And also, they actually organize us and ask us, what are the problems you are facing, or is there any problem or something, tell us so that the elders there, maybe they can help you or something, you know? So we were being encouraged by the elders, and when we go from the church they shake our hands, and maybe some of them with a car they can come and drop you at the house. You see? And sometimes even they will tell you maybe you don't have a Bible or something, they can help you with a Bible. Yeah, so they were good, you know, they were doing a lot of things, I appreciate very much when I went there.

(Vincent, 4 May 2006)

Vincent's attraction to each church is due largely to his perception of its capacity to contribute to his practical needs. At Shiloh and the AG this is in a general sense: 'if you have any difficulties the pastor can help you', while he attends Liberty in search of solutions to very specific issues and, after two years, leaves the church as his problems remain unsolved. The appeal of the AG lies in its practical instruction on how to live, in the generosity of its elders and in the encouragement and recognition he receives. Membership of the church is an opportunity to move upwards in society: 'the people want me to be somebody'. To Vincent, then, religion is essentially practical, a strategy to resolve his problems.

As well as switching between different churches serially, it is not uncommon for people who consider themselves members of one church to attend meetings of another church while continuing to attend their own church services as normal. Usually the extra meetings attended are prayer meetings, the motivation being to obtain prayer for a specific problem. The reputation of certain churches (or often certain individuals within churches) as being particularly powerful in bringing about solutions and miracles is a key reason for movement and overlapping between churches. Movement between churches within the Pentecostal sphere is more likely than between Pentecostal churches and AICs, since the former (such as the AG) generally do not recognize most AICs (such as the MDCC) as 'properly' Christian.

However, the choice of church does not depend solely on the reputation of the church in question and the perceived effectiveness of its prayers; the circumstances and relationships of the person concerned are also determining factors. Where people have strong relationships with others in different churches denominational boundaries may be overstepped. For example, while Abena, the sister-in-law and household member of the MDCC pastor, is a member of the Church of Pentecost, she also participates in some of the Musama meetings, usually when they involve an issue directly concerning her. So, for example, she attended the Sunday morning MDCC service on the occasion of the church presentation of her granddaughter, whose mother (Abena's daughter) is an MDCC member. She also, before leaving with her daughter and grandchildren on a visit to relatives on the far side of the country, took part with the other members of the household in Musama prayers for their protection during the journey; and she participates in occasional impromptu household prayer meetings called by the pastor (usually as a response to dreams and visions) in the middle of the night. The latter two events would appear to be partly a question of submission to the authority of her sister-in-law, who is her senior in age and head of the household, as well as possessing additional authority in her role as a church leader. Conversely Abena's daughter, Beth, when facing relationship difficulties sometimes attended a weekday prayer meeting at the Church of Pentecost in Takoradi. The choice of church is connected to her mother's positive experience of prayer there through which she was healed from a serious illness, leading to her enrolling as a member. Beth's position as an MDCC member did not pose any problems to her seeking additional prayer elsewhere; while the AG church does not consider most AICs as Christian, the MDCC is far broader in its recognition of other denominations.

Continuity between 'religious' and 'secular': healthcare strategies

Gifford (2004) argues that the new charismatic churches in urban centres privilege spiritual over structural or scientific causes and explanations of problems, thereby encouraging people to spend hours and days at meetings praying for wealth instead of working, or to forsake medical treatment for divine healing. This, however, does not seem to be the case in either of the churches in this study. The AG church runs two hospitals in the north of Ghana and includes medical outreach as part of its mission statement which is to minister to all the needs of the people (Ghana Evangel, 2006, p. 3), combining prayer and medical treatment rather than substituting the former for the latter. This is reflected in the approaches of AG members to illness, as illustrated in the following testimony given by a female teacher during a church service:

> Last Friday was my daughter's second birthday. That evening at six o'clock I was in the kitchen preparing supper and something struck me to go into the house. I went there and found the child with a bottle of kerosene that I had hidden somewhere – she had taken the kerosene to rub all over her body and I suspected that she might have drunk some. So I rushed her to the hospital and she was given some medicine and vomited everything up. I want to thank God for saving my daughter.
>
> (Natasha, 14 May 2008)

Here, God is given overall credit for the protection and recovery of the child, both alerting her mother to the danger and working through hospital staff to treat her. However, at the same time medical treatment is considered essential for resolving what is interpreted as a medical issue. Two levels of operation can therefore be identified: while God has ultimate control over everything and can intervene as desired, other systems work within those parameters and must be complied with. Prayer is thus an ongoing essential in all areas of life, but it does not preclude the necessity also to interpret and approach problems in non-religious terms. The quality of being 'religious' or 'non-religious' is itself not distinguished: there is no sense that in turning to 'secular' medicine people are any less seeking help from God, rather there is a sense that medicine acts as a medium through which God's power is manifested. This notion is not, of course, unique to this particular church or to Ghana. It can be found in other religions and societies, including in developed countries.

According to Baëta, MDCC members may not consult either Western-trained doctors or African herbalists and 'are not allowed the use of any drugs whatsoever, and may be treated in the hospital only for accidents involving some abrupt break of a bodily organ' (1962: 54). Although divine healing remains one of the key concerns of the church, this is no longer to the exclusion of other forms of medicine – indeed, the Ndwumizili pastor is a trained nurse and runs the village pharmacy. She is visited by people with all kinds of health issues and her treatments include biomedical tablets and injections and advice on diet and lifestyle as well as prayer and religious (specifically Musama) rituals. Prayer is not always an intrinsic or necessary component of her treatments: her most common prescriptions include drugs such as chloroquin, paracetamol, imodium and vitamin supplements, and she frequently refers cases beyond her scope to a hospital in one of the neighbouring towns. On occasion she will launch into a tirade against people who do not seek appropriate medical attention, for example when a member of her church whose baby was burned in a fire waited four days for the pastor to return from travelling instead of taking him to hospital. The reason for the delay was not about preferences between particular types of medical approaches or remedies, but money: hospital visits must be paid for, while the pastor can be persuaded to give treatment on credit. Thus, as in the AG church, 'modern' medical science is not rejected, while religious practices are also employed. Both MDCC prayers and 'modern' medicines (as well as local herbal remedies) are seen as elements of a wider range of treatments, to be used as appropriate according to the diagnosis.

As the church is not the only abode of power, in addressing their problems and in seeking protection and blessings people may also turn elsewhere. They draw not only on modern powers such as biomedicine and education, and on support from institutions such as the family, but they may seek help from other spiritual powers, through juju practitioners, *mallams* (Islamic holy men), herbalists and fetish priest(esse)s[3] (areas beyond the scope of this paper). Whether seeking help from powers within or outside the church spectrum, the employment of multiple strategies does not necessarily take place in a linear fashion: people do not simply move through a number of different social or religious discourses discarding each until they find one that works. They may do this with respect to the powers themselves, for example trying a range of churches in search of one with effective solutions or neglecting deities that do not deliver results; however, these powers do not necessarily represent whole religions or philosophies. Rather, because the world is conceptualized in terms of powers rather than belief or thought systems, it is quite possible – and common – for people to combine different 'discourses', drawing on them simultaneously in a pragmatic fashion, without any sense of contradiction.

These observations reflect literature from interpretive medical anthropology, which demonstrates that in seeking effective healthcare people tend to act pragmatically, drawing on and combining different socio-medical 'systems' as they exercise different options without necessarily accepting their ideological bases. As Lock and Nichter note, 'patients are, almost without exception, pragmatic, and see nothing inconsistent about liberally combining different forms of therapy in their quest for restored health' (2002, p. 4). Brodwin, researching medical pluralism in Haiti, asserts that '[r]eligious affiliation emerges from people's response to bodily and social affliction, not their cognitive acceptance of a set of cosmic principles or abstract propositions about good and evil' (1996, p. 18). He demonstrates how people continually renegotiate their position in relation to different religions and medical systems in the face of the trials and challenges they encounter in everyday life and at crisis points. This happens on an intrinsically pragmatic level which allows them to move between systems without assuming their ideological bases. A similar observation is emphasized by Last in his study of medical cultures in Hausaland: 'People do not, in my experience, face intellectual problems in embarking on the appropriate method of treatment...there are many more pressing, practical problems with which to cope' (Last, 1992, p. 403).

Regarding the study of religion in Africa, Brenner asserts that although in academic studies religious knowledge is privileged over religious participation, in reality in Africa participation precedes and is more widely available than knowledge, which is often esoteric, possessed exclusively by priests and elders. Most people, then, do not start with an awareness of a coherent cosmology – whether or not one exists – and act accordingly; rather, they start with practice, participating in ritual or reacting to situations, and their conceptual universe is constructed around this. There is thus not only space for a great deal of variation and eclecticism in explanations of events and occurrences, but also considerable scope for religious and intellectual creativity (Brenner, 1989, p. 91). On the other hand, to privilege embodied participation and practice too much can imply that people are not capable of cognitive thought and risks endorsing racial stereotypes of Africans as primarily bodily beings while Westerners are intellectually oriented. Mind and body are, of course, fundamentally linked and even if practice precedes and is privileged over knowledge, it only occurs within a normative context informed by knowledge, however subconsciously. In the context of churches where knowledge is not entirely esoteric and is passed on and reconstructed through sermons, discussions and Bible studies, it is important not to ignore doctrines and teachings as well as how these are received, interpreted and acted upon.

Conclusion

Understanding the world as a continuous landscape occupied by a range of powers – both spirits and humans – rather than seeing those powers as constructs belonging uniquely to separate socio-religious discourses, provides a different perspective on the ways people act and relate in Ghanaian society. It also challenges some of the categories normally assumed by Western social science. Firstly, the boundaries between different church denominations are called into question. A shift in focus from the churches to their members reveals that the boundaries between specific churches, between denominations and between religious traditions are not fixed. People are likely to switch between and combine strategies from all of these, marshalling different powers, as they seek to protect themselves from harm and acquire blessings. The implication is that in order to gain a full picture of how religion is played out in the lives of Ghanaians (and potentially more widely in Africa), studies need to move beyond their church or organizational orientation and become more person-based. Secondly, the findings of this research suggest the need to question some of the dualistic assumptions that lie beneath much social science analysis of religion. In particular, the opposition between the 'religious' and the 'secular' should be recognized and challenged. A development discourse which, based on Enlightenment principles, seeks to overcome the irrational with the rational, needs to be qualified with the awareness that many people in developing (and developed) countries do not share this objective. Rather than renouncing 'irrational' religious beliefs, values and actions and replacing them with 'rational' secular traditions as modernity progresses, people in Ndwumizili construct personalized cosmologies on a practical basis, drawing on the different discourses available to them. These include discourses which are often labelled as 'religious', such as juju, witchcraft, communication with ancestors and various types of Christianity, although the people concerned may not consider them as such. On the other hand, other 'secular' discourses which are similarly drawn on are also considered to have a spiritual element. The approach suggested by this study to enable us to move beyond the religious/secular opposition is, through careful ethnographic research, to expose the inadequacies of such conceptual frameworks and to develop others which represent more accurately the perspective of the people in question.

If development agencies want to be genuinely actor-oriented, seriously taking on board what people say and think, when they engage with religion they must do so in a way that is meaningful and relevant to the everyday reality of people's lives. As well as working to form partnerships with organized religious bodies and institutions they must be aware of the fluidity, eclecticism, diversity and multiplicity that are concealed by the institutional landscapes. Development theoreticians and practitioners must take care in treating 'faith communities' as discrete blocks and recognize that for many people religion consists mainly of practice rather than abstract theology. Members of 'faith communities' may not only switch between different groups, but may interpret and draw on the discourses of such groups simultaneously with other discourses as they pragmatically manage their daily lives. It is therefore simplistic and probably misleading to assume that messages sent out by churches alone either shape or represent the values and actions of their congregations: each church – or church leader – is just one power among many. Furthermore, churches are not always or perfectly altruistic: although they may provide networks and resources, these resources are likely to be shared primarily among the members of each denomination (or among their leaders) rather than distributed externally and equally to all. Indeed, churches may be more concerned with making money (and converts) and increasing their own status and the prosperity of their leaders and members than with community development. Churches are also not free from their cultural context. Fear of malicious or self-interested intent on the part of others, constructed particularly in the language of witchcraft and juju, results in distrust, suspicion and secrecy even within church congregations. While they can go some way in providing new forms of identity and belonging, even churches such as the AG, which claims identity with a global community of Pentecostals, are still largely constrained within local cosmologies and power relations.

Finally, development actors should be aware of their own position in people's worldviews. Development understood as social or economic change, or as poverty reduction, capacity building, empowerment or freedom can offer potential means of marshalling the powers. However, development is not a neutral force: it is often embodied in the form of agencies and workers, and it is ambivalent, bringing both blessings and harm. Development interventions and actors are not free from being interpreted through religious discourses. They represent powers within the cosmological landscape of the residents of Ndwumizili. As such, as well as endeavouring to impose their own agenda, they can be engaged with and marshalled as people seek to negotiate those powers in their everyday lives. The framework and call for detailed ethnographic work set out in this chapter entails recognizing that while development theoreticians and practitioners may consider religion in an instrumental fashion – asking what role religion can play in development – this can work both ways. Development can also be employed instrumentally as people draw on religious discourses to negotiate the world, using development agencies and workers present in their community to maximize their blessings and protect them from harm. In one sense, of course, this is exactly what development is for: to be used by poor people for their benefit. However, it may be used and interpreted in ways other than it intends: programmes may be drawn on with practically no acceptance of their ideological bases; benefits may be construed as blessings from God or as originating with evil powers; workers and beneficiaries may be implicated in relationships entailing bias, obligation and suspicion. The crucial question, therefore, not only regards the role of religion in development, but also the implication of development within religious imaginaries.

Notes

1. The name of the village and the names of informants have been changed. This analysis draws on ethnographic data collected over a period of 18 months in 2005–06 as part of doctoral research sponsored by the ESRC.
2. For example, 'Check fake and illegal churches', *Mirror*, 16 September 2006; 'Pastor steals musician's money', *Spectator*, 6 January 2007.
3. Secondary deities, or 'small gods', are commonly referred to as 'fetishes' in Ghana.

References

Baëta, C. G. (1962), *Prophetism in Ghana: A Study of Some 'Spiritual' Churches*, London: SCM Press.

Bastina, M. (1993), ' "Bloodhounds Who Have No Friends": Witchcraft and locality in the Nigerian Popular Press' in J. Comaroff & J. Comaroff (eds), *Modernity and Its Malcontents: Ritual and Power in Postcolonial Africa*, Chicago: Chicago University Press, pp. 129–66.

Barrett, D. B. (2001), *World Christian Encyclopedia: A Comparative Survey of Churches and Religions in the Modern World*, Oxford: Oxford University Press.

Belshaw, D., Calderisi, R. & Sugden, C. (eds) (2001), *Faith in Development: Partnership between the World Bank and the Churches of Africa*, Oxford: Regnum/World Bank.

Brenner, L. (1989), ' "Religious" Discourses In and About Africa', in Barber, K. and Farias de Moraes, P. (eds) *Discourse and its Disguises: The Interpretation of African Oral Texts*, Birmingham: Centre for West African Studies, University of Birmingham, pp. 87–105.

Brodwin, P. (1996), *Medicine and Morality in Haiti: The Contest for Healing Power*, Cambridge: Cambridge University Press.

Cameron, H. (2003), 'The Decline of the Church in England as a Local Member-ship Organization: Predicting the Nature of Civil Society in 2050', in Davie, G., Heelas, P. & Woodhead, L. (eds) *Predicting Religion: Christian, Secular and Alternative Futures*, Aldershot: Ashgate, ch. 9, pp. 109–19.

Ciekawy, D. & Geschiere, P. (1998), 'Containing Witchcraft: Conflicting Scenarios in Postcolonial Africa' *African Studies Review* 41 (3), pp. 1–14.

Clarke, G. (2006), 'Faith Matters: Faith-Based Organizations, Civil Society and International Development' *Journal of International Development* 18, pp. 835–48.

Comaroff, J. & Comaroff, J. (1993), 'Introduction', in Comaroff, J. & Comaroff, J. (eds) *Modernity and Its Malcontents: Ritual and Power in Postcolonial Africa*, Chicago: University of Chicago Press, pp. xi–xxxvii.

Ellis, S. & ter Haar, G. (2004), *Worlds of Power: Religious Thought and Political Practice in Africa*, London: Hurst and Co.

Fasholé-Luke, E., Gray, R., Hastings, A. & Tasie, G. (eds) (1978), *Christianity in Independent Africa*, London: Collings.

Fernandez, J. W. (1978), 'African Religious Movements' *Annual Review of Anthropology* 7, pp. 195–234.

Foli, R. (2001), *The Church in Ghana Today*, Accra: Methodist Book Depot Ghana.

Gifford, P. (2004), *Ghana's New Christianity: Pentecostalism in a Globalizing African Economy*, Bloomington and Indianapolis: Indiana University Press.

Hackett, R. I. J. (2003), 'Discourses of Demonization in Africa and Beyond' *Diogenes* 50 (3), pp. 61–75.

Haynes, J. (1993), *Religion in Third World Politics*, Buckingham: Open University Press.

Heelas, P. & Woodhead, L. (2005), *The Spiritual Revolution: Why Religion is Giving Way to Spirituality*, Oxford: Blackwell.

Larbi, E. K. (2001), 'The Nature of Continuity and Discontinuity of the Ghanaian Pentecostal Concept of Salvation in African Cosmology' *Cyberjournal for Pentecostal/Charismatic Research* 10.

Last, M. (1992), 'The Importance of Knowing about Not Knowing: Observations from Hausaland', in Feierman, S., Janzen, J. M. (eds) *The Social Basis of Health and Healing in Africa*, Berkeley/Los Angeles/London: University of California Press, ch. 16, pp. 393–406.

Lock, M. & Nichter, M. (2002), 'Introduction: From Documenting Medical Pluralism to Critical Interpretations of Globalized Health Knowledge, Policies, and Practices', in Nichter, M. & Lock, M. (eds) *New Horizons in Medical Anthropology*, London: Routledge, ch. 1, pp. 1–34.

Marshall, K. & van Saanen, M. (2007), *Development and Faith: Where Mind, Heart and Soul Work Together*, Washington, D.C.: World Bank.

Martin, D. (2001), *Pentecostalism: The World Their Parish*, Oxford: Blackwell.

Meyer, B. (1998), ' "Make a Complete Break with the Past": Memory and Post-Colonial Modernity in Ghanaian Pentecostalist Discourse' *Journal of Religion in Africa* 28, pp. 316–49.

Meyer, B. (2004), 'Christianity in Africa: From African Independent to Pentecostal-Charismatic Churches' *Annual Review of Anthropology* 33, pp. 447–74.

Moore, H. L. & Sanders, T. (2001), *Magical Interpretations, Material Realities: Modernity, Witchcraft and the Occult in Postcolonial Africa*, London: Routledge.

Petito, F. & Hatzopoulos, P. (eds) (2003), *Religion in International Relations: The Return from Exile*, Basingstoke: Palgrave Macmillan.

Rakodi, C. (2007), 'Understanding the Roles of Religions in Development: The Approach of the RaD Programme, Religions and Development Working Paper 9', International Development Department: University of Birmingham.

Sackey, B. (2001), 'Charismatics, Independents, and Missions: Church Proliferation in Ghana' *Culture and Religion* 2 (1), pp. 41–59.

ter Haar, G. & Ellis, S. (2006), 'The Role of Religion in Development: Towards a New Relationship between the European Union and Africa' *European Journal of Development Research* 18 (3), pp. 351–67.

Thomas, S. M. (2003), 'Taking Religious and Cultural Pluralism Seriously: The Global Resurgence of Religion and the Transformation of International Society', in Petito, F. & Hatzopoulos, P. (eds) *Religion in International Relations: The Return from Exile*, Basingstoke: Palgrave Macmillan, ch. 1, pp. 21–53.

Tyndale, W. R. (2006), *Visions of Development: Faith-Based Initiatives*, Aldershot: Ashgate.

Ver Beek, K. A. (2000), 'Spirituality: A Development Taboo' *Development in Practice* 10 (1), pp. 31–43.

10
Sacred Struggles: The World Council of Churches and the HIV Epidemic in Africa

Ezra Chitando

Introduction

The interface between religion and development has elicited varying scholarly responses. Given the elasticity of the concepts of religion and development, there continues to be debate on whether and how religion affects development. This chapter focuses on the World Council of Churches (WCC) and its project, the Ecumenical HIV and AIDS Initiative in Africa (EHAIA), to reflect on the interplay between religion and development. It contends that while the WCC has sought to contribute to development by mitigating the effects of HIV and AIDS in Africa, its status as an external development agency militates against its efforts. The WCC has lofty ideas and ideals, and encourages its member churches to meet them, but it does not have the mechanism to enforce its vision. In short, it relies on the goodwill of its member churches to bring its vision to fruition.

In their publication, *Religion in Development, Rewriting the Secular Script*, Séverine Deneulin and Masooda Bano (2009, p. 4) assert that 'there is no separation between religion and development'. This chapter corroborates this understanding by highlighting how the WCC has regarded its HIV intervention as part of its mission. However, the chapter notes that as an international ecumenical body, the WCC has strategic advantages and serious limitations in its efforts to effect 'development' at national and local levels.

Background: HIV and the challenge of development in Africa

The Human Immunodeficiency Virus (HIV) and the Acquired Immunodeficiency Syndrome (AIDS) epidemic have left behind a trail of destruction in most parts of sub-Saharan Africa. Southern Africa has been particularly affected, with AIDS becoming the leading cause of death in the region (Weinreich and Benn, 2004, p. 8). A history of the epidemic shows that it has conquered most parts of black Africa, leaving a trail of death and destruction (Illife, 2006). Communities have been ripped apart, while the number of orphaned children has grown phenomenally.

Various religious groups in Africa (including the different Christian and Muslim groups) and abroad have responded differently to the HIV epidemic. In most instances, they have followed the scheme of denial-condemnation-active engagement. In the initial instance of denial (between the mid-1980s and 2000), religious communities either completely denied the existence of the epidemic or minimized its potential impact. In the second phase, religious groups averred that the epidemic had eschatological significance. Especially within Christian circles, the HIV epidemic came to be regarded as God's judgment on an apostate generation. Admittedly, not all churches reached this verdict. Nonetheless, it proved to be the dominant paradigm. In the third phase, religious groups have sought to provide effective responses to the epidemic. They have mobilized communities to resist stigma and discrimination of people living with HIV. They have also called for effective leadership in various fields to ensure that the struggle against HIV and AIDS is a successful one. There is now a growing realization that the HIV epidemic transcends the issues of health and well-being. HIV raises key questions about the role of churches in politics and development. A publication by the World Bank asserts that:

> The challenges of fighting the HIV/AIDS pandemic go far beyond the specific issues of health and medical care. HIV/AIDS involves every facet of the development agenda, and thus needs to engage the widest possible variety of partners – private businesses, NGOs, community organizations, government, and perhaps most directly, faith communities.
>
> (Marshall and Keough, 2004, p. 135)

This chapter selects the WCC's engagement with HIV in Africa as a case study to understand the religion-development dynamic. As the largest ecumenical body in the world, the WCC's response to HIV provides valuable insights into the relationship between religion and development. The 'global' character of the WCC enables it to access public spaces in Africa that single denominations would (and have struggled) to access. With its head office in Geneva, Switzerland, alongside many international development agencies, the WCC aims to give religion a platform within international development decision-making.

By responding to the challenge of HIV as a consortium of churches and ecumenical bodies, the WCC offers an intriguing dimension to the notion of 'religion'. Whereas most researches on religion and development focus on specific denominations, the WCC brings together churches that have different theological positions. However, by placing emphasis on the notion of Christianity as providing 'abundant life' to all, that is, ensuring that the life before death is as comfortable as it can be, the WCC has been able to bring together the various churches and ecumenical organizations to collaborate in the quest for development.

Although the WCC would not regard itself as a 'development agency', this chapter argues that its HIV interventions can be regarded as 'development work'. Its overall aim of developing 'HIV competent churches' (Parry, 2008) is couched in the language of development. The WCC's strategies to tackle HIV include gender empowerment, leadership training and community mobilization. These strategies seek to transform communities that have been adversely affected by HIV into vibrant, self-reliant communities. In this chapter, I understand development to represent the transition from a state of hopelessness and despair to one of feeling equipped to shape one's destiny. Development is linked to empowerment: communities that are convinced that they change their plight are well on their way to development. I will argue that the WCC has sought to contribute to development by challenging the paralysis that most African communities felt in relation to HIV. It has endeavoured to energize African communities to face the epidemic with creativity and confidence that a prosperous future is possible.

Reflecting on the WCC's interventions to address HIV in Africa poses a major challenge to me as a member of the WCC employed as a Theology Consultant on HIV and AIDS. This raises the insider-outsider problem in research and writing (Adogame and Chitando, 2005). On the one hand, being an 'insider' facilitates access to the ideas and processes involved in the WCC's HIV interventions in Africa. This is clearly an advantage. On the other hand, there is a danger that I might uncritically celebrate the WCC's achievements as I am closely related to its HIV work. Furthermore, there is the challenge that, in the name of seeking to appear more 'objective', I might understate the significance of WCC interventions.

Reflexivity is important in research and writing. It helps the scholar to be aware of the pitfalls and prepares him or her to become more careful. I have sought to provide a balanced account of the WCC's HIV work in Africa. I am painfully aware of the reality that 'objectivity' is difficult to attain. Nonetheless, I have endeavoured to highlight the strengths and weaknesses of the WCC's engagement with HIV in Africa as I perceive them.

Numb and overwhelmed: earlier responses to the HIV epidemic by African churches

The mid-1980s saw many African governments acknowledging their first HIV cases. As the disease had initially been associated with the gay community in the United States of America, there was considerable apprehension and panic. In addition, most African governments and intellectuals were forced on the defensive as they sought to counter racist arguments that portrayed Africa as the source of the disease (Chirimuuta & Chirimuuta, 1987). Furthermore, there were fears that tourism would suffer if governments were too quick to admit having many people living with HIV.

There were two dominant responses from churches in Africa. On the one hand, churches did not take the HIV epidemic seriously. It was presented as a challenge for 'outsiders', that is, those who were not members of the churches. The self-understanding of the church as 'the body of the saved' allowed it to regard HIV as an issue for 'others'. Christians, as people who upheld the commandments of God, were safe, at least according to this line of thinking. It was believed that the emerging disease threatened only those who 'were not covered by the blood of Christ'. Some Pentecostal churches in particular popularized this response to the epidemic. Similar discourses emerged among Muslims. Nadine Beckmann (2009) illustrates how Muslims in Zanzibar frame the discourse on AIDS in terms of narratives of decline. They look back to the past for moral values that, ostensibly, starve off HIV.

On the other hand, the churches fuelled stigma and discrimination. By framing HIV discourses in moralistic tones, churches presented people living with HIV as 'sinners' who failed to uphold Christian standards of ethical behaviour. Preachers presented HIV as a sign heralding the 'end of times', and people living with HIV as living testimonies to the moral failings characterizing a fallen generation. In the early phase of the epidemic, churches were generally harsh on people living with HIV. They were unable to 'think outside the box' in the face of AIDS-related stigma (Paterson, 2005).

In order to understand the churches' struggle with HIV, there is need to appreciate the fact that churches have to deal with issues of sexual education, morality and theological interpretations. In tropical Africa, heterosexual activity constitutes by far the most dominant mode of transmission for HIV. This has posed a major problem as churches have historically found it difficult to communicate effectively on the theme of sexuality (Schmid, 2007). The dominant trend has been to associate sexuality with temptation, sin and human weakness. The inherited flesh-spirit dichotomy has not empowered churches to address human sexuality openly and effectively. This led in general to African churches adopting a negative attitude towards people living with HIV.

It is correct to state that during the first decade of the HIV epidemic African churches struggled to provide adequate practical and theological responses. This was mainly due to the fact that the dominant theology did not galvanize the churches into action. Theological mediocrity prevented the churches from interpreting the HIV epidemic in inspired ways. I define theological mediocrity as the failure to 'read the signs of the times' and make churches relevant to their socio-economic and political contexts. Although many churches and their members provided care and support to the infected and the affected, their theological premises were still inadequate. Many continued to associate HIV with divine punishment. It was felt that people living with HIV were being punished for their (assumed) laxity in sexual matters.

One of the most debilitating debates that prevented the churches from providing effective responses to HIV and AIDS was the issue of condoms. Valuable time was wasted as some church leaders attacked the use of contraceptives instead of confronting the virus. Many believed (and some still do) that churches had 'no business' talking about condoms. They regard Christians as 'saved' people who are beyond negotiating their sexuality on a regular basis. In particular, many Pentecostal/Charismatic Christians have opposed the use of condoms. They have sought to place exclusive emphasis on abstinence and faithfulness, in line with what they regarded as 'old time religion'. Others sought to question the effectiveness of condoms. The aversion to sexuality described above, and rigid insistence on purity prevented churches from adopting a more humane approach. Writing from within the Catholic Church, Charles Ryan describes the challenge:

> Could we imagine ourselves as a Church saying: 'Sexual activity outside marriage is harmful and unsafe, and we wish you did not engage in it, but if you insist on being sexually active outside marriage, please, at least use a condom'? But, if not, why not? In spite of expressed doubts about the efficacy of condoms to prevent the

transmission of HIV status, there is no doubt that using a condom does afford a degree of protection against this deadly condition. The technical reality is that sex with a condom is safer than sex without a condom – all other things being equal.

(Ryan, 2003, p. 9)

It was this earlier approach that led many activists to regard churches and faith-based organizations as part of the problem, rather than as being part of the solution to the challenge of HIV and AIDS. Churches were portrayed as unreasonable, dogmatic and proclaiming death at a time when the world needed a message of healing and hope. As part of the stigma, some Christian ministers of religion refused to conduct funeral services for individuals who had died of AIDS. Fiery sermons condemning people living with HIV had the effect of presenting the churches as detached and uncaring.

It must be acknowledged that despite the overall negative approach towards the HIV epidemic in the 1980s and 1990s, there were instances when churches and their members acted prophetically. Some churches and their leaders were willing to journey with people living with HIV at a time when stigma and discrimination were at their highest. Churches also provided hospitals and other medical facilities that eased the suffering of many people living with AIDS. Many individual Christians also overcame the numbness that characterized the institutional church and demonstrated love and compassion towards people living with HIV. These values have been an integral part of the Christian faith. For example Cimperman stated that:

Well before the emergence of AIDS, the Church worked across the globe in health care, education, and social services. Through clinics and hospitals, elementary and secondary schools, colleges and professional schools, orphanages, homeless shelters, counseling services, food distribution centers, and much more, the Gospel mandate to love and serve one another is incarnated wherever the need arises. Direct service with and among the people gives the Church a particular competence in speaking to policy issues that directly affect the people of God.

(Cimperman, 2005, p. 15)

The church's struggle to respond effectively to HIV in the early years of the epidemic can be attributed to its tendency to regard its mission in very narrow terms. Once the discourse had been framed within the context of morality and the salvation of the soul, it became extremely difficult for the different denominations to regard HIV as an integral part of the mission of the church. Secondly, the tendency to regard 'development work' as falling outside the African church's sphere of influence paralysed religious organizations.

Islam in Africa encountering HIV: an overview of the paralysis

In case one might conclude that only the Christians were overwhelmed by HIV in the early phase of the epidemic, it is instructive to note that Muslims also struggled to provide effective responses. In an introduction to their edited volume, *Islam and AIDS: Between Scorn, Pity and Justice*, Farid Esack and Sarah Chiddy (2009, p. 2) write:

> Like many others, Muslim societies, and those approaching the pandemic as Muslims, have responded to AIDS in a clearly discernable pattern: first ignorance and denial, then scorn and pity.

As was the case with the Christians, Muslim organizations tended to adopt a very narrow and moralistic framework when facing HIV. Adopting a fundamentalist position, many argued that 'good Muslims' would not get infected with HIV. They regarded HIV as a signifier of the moral status of the infected individual. Furthermore, there is an assumption that HIV affects only those outside the *ummah* (Islamic community of faith). In *Scaling up Effective Partnerships: A Guide to Working with Faith-based Organisations in the Response to HIV and AIDS*, Steven Lux and Kristine Greenaway make the following observation:

> The primary obstacle to collaboration is denial that HIV and AIDS already affects Muslims and has the potential to affect them as much as any other community. A second obstacle is the belief that moral, upright Islamic behaviour is the only answer to this problem, combined with the refusal to look at the larger structural and systemic issues of inequality, poverty and injustice that lie at the roots of the spread of the disease.
>
> (Lux and Greenaway, 2006, p. 81)

The initial Muslim paralysis in engaging HIV, like that of the Christians, was informed by theological rigidity. The dominant theological paradigm was to understand HIV within the context of personal sin. Very little or no attention was paid to structural sin, or what Lux and Greenaway (2006, p. 81) in the citation above call 'the larger structural and systematic issues'. Religious leaders were either unable or unwilling to locate HIV within the larger context of poverty and gender inequality.

In general, civil society actors regarded religious leaders as a stumbling block to development. In particular, HIV and AIDS service-care organizations felt that religious leaders were spending more time and energy fighting the condom and not the virus. Religion was experienced and understood as a barrier to development. This was unfortunate as precious time was wasted in these debates. The problem was so important that it called for religious leaders and churches in Africa to regard the HIV epidemic as part of their core business. It was with the WCC's intervention that churches in Africa either began to, or increased, their engagement with HIV. The chapter turns to this process of the WCC's mobilization of the churches in Africa. Initially, however, it highlights the WCC's prophetic ministry in general and in combating racism in Africa in particular.

Consistent prophetic action: an overview of WCC activities in sub-Saharan Africa

Although the WCC did not undertake radical actions aimed at addressing the HIV epidemic during the 1980s and 1990s, it was consistent in its pronouncements on the need to fight stigma and discrimination. The WCC is undoubtedly the world's largest ecumenical body, and has had a distinguished history in the struggle for justice. While its member churches enjoy autonomy, the WCC has sought to promote theological maturity among these churches. Theological maturity is understood as the capacity to understand the church's mission in the contemporary world. It is the recognition that the church is not just concerned with heavenly matters, but also with the here and now. Theological maturity is indicated by the church's willingness to address contemporary socio-economic and political issues. The WCC has continued to expand and to respond to contemporary challenges. Thus:

> In 2005, the WCC had a membership of 348 member churches which together claimed 592 million Christian members in more than 120 countries. WCC member churches include nearly all the Eastern and Oriental Orthodox churches; Anglicans; diverse Protestant churches, including Reformed, Lutheran, Methodist, and Baptist, and a broad

representation of united and independent churches. While most of the WCC's founding churches were European and North American, today the majority are in Africa, Asia, the Caribbean, Latin America, the Middle East and the Pacific.

(van Beek, 2006, p. 3)

The WCC has a long history in promoting public action in African societies. It is important to recall that the organization had been an integral part of the struggle against racism and apartheid in Southern Africa. Through the Programme to Combat Racism (PCR) that was established in 1969 following the Central Committee's mandate in 1968, the WCC provided material and moral support to liberation movements in Southern Africa. Liberation movements in Angola, Namibia, South Africa and Zimbabwe received grants from the PCR. This proved controversial as some churches in the Global North feared that through the PCR, the WCC was promoting communism. However, liberation movements in Southern Africa insisted that racism and apartheid were sins, and Christians had a responsibility to resist them (Banana, 1996).

The importance of the PCR to the struggle against racism was clearly demonstrated when Nelson Mandela, the anti-apartheid icon and then President of South Africa, visited the WCC Eighth Assembly in Harare, Zimbabwe, in December 1998. Mandela thanked the WCC for having played a major role in his country's struggle against apartheid. Where some had feared 'mixing religion with politics', the WCC had taken a clear stance in opposing apartheid. During the same occasion, Robert Mugabe, President of Zimbabwe, also thanked the WCC for supporting liberation movements in the region.

Alongside the PCR, the WCC has supported many social justice issues. These include the struggle against violence, the marginalization of women, the cause of the Dalits, ecological concerns and others. Although the WCC does not have a specific theology, representing as it does a multiplicity of churches, it is clear that liberation theology has tended to guide most of its staff. There is a firm commitment towards standing with those at the margins, and working towards social justice. Following liberation theology's 'preferential option for the poor' (Núñez & Taylor, 1996), the WCC has played an important advocacy role. It has supported debt cancellation for developing countries, building on the biblical principle of the jubilee. It has called for economic justice, and for promoting the interests of developing countries in a world dominated by a few rich nations. The WCC has charged that 'development' that leaves the majority of the world's citizens poorer is unsustainable.

In its operation, the WCC joins hands with progressive organizations to challenge oppressive economic and political systems. At the heart of its theology is the contention that 'mission' is not confined to preaching within the boundaries of the church buildings. Rather, the WCC regards organizations working for justice and social transformation as undertaking mission. As a result, it has forged effective partnerships with organizations like Jubilee 2000 that works for debt cancellation and the Ecumenical Advocacy Alliance that has played a major advocacy role.

It is within the context of the WCC's prophetic ministry that its response to the HIV epidemic must be located. At a time when most Christians adopted a negative attitude towards the epidemic, the WCC demonstrated appreciable theological maturity. As early as 1987, the WCC executive committee drew attention to the need for churches to provide effective responses to HIV and AIDS. It called for urgent responses in the areas of pastoral care, education for prevention and social ministry (WCC, 1997, p. 96). This was significant as the dominant trend had to been to condemn people living with HIV. The statement adopted by the WCC Central Committee in September 1996 called for a shift in the theology of the churches, and paid attention to long-term causes and factors encouraging the spread of HIV and AIDS. This was equally significant as there had been a preoccupation with individual morality, as indicated in the previous sections.

The shift towards greater engagement with HIV was a result of collaboration between the churches of Africa, donor agencies and ecumenical bodies. The churches of Africa had reached the realization that HIV was not only 'out there', but was very much within the churches themselves. Donor agencies felt that they, the churches, had to contribute as they were an integral part of the world.

It is vital to point out that the WCC's prophetic action has not always been appreciated by some of its members. For example, the PCR was highly divisive (Banana, 1996). Some members felt that the WCC's liberation theology was 'non-Christian' and was inspired by Marxism. Whereas there is unanimity regarding the appropriateness of 'preaching the gospel', engaging in civic action has been more contentious.

The ecumenical HIV/AIDS initiative in Africa: striving for AIDS-competent Churches

In line with the WCC's vision of the role of the churches in responding to the HIV epidemic, a number of consultative meetings were held in the different regions of Africa. These took place in 2001 in Uganda, South

Africa and Senegal. African Church leaders, ecumenical bodies, national councils of churches and related non-governmental organizations participated in these meetings (Plan of Action, 2002, p. 2). The major focus was on how to set up a unit that would enhance the African churches' response to HIV and AIDS. The WCC facilitated a 'Global Consultation on the Ecumenical Response to the Challenge of HIV/AIDS in Africa', in Nairobi, Kenya, 25–8 November 2001. This laid the foundation for the emergence of EHAIA in 2002.

Operating within the WCC, EHAIA has experienced remarkable growth. It has a project coordinator and an administrative assistant based in Geneva, Switzerland. All other members of staff are based in Africa. These include the regional coordinators for Southern, Eastern, Central, West and Lusophone Africa. There are two theology consultants for the Anglophone and Francophone regions. Regional Reference Groups guide the regional coordinators, while the International Reference Group provides guidance to the project as a whole. In the following section, I outline some of the major areas of collaboration between EHAIA and African churches in the struggle against the HIV epidemic.

This structure has been welcomed by African churches as it gives them a voice and sense of ownership. The Regional Reference Groups (RRGs) are made up of representatives of various churches and ecumenical bodies within the particular region. They work with the regional coordinators and theology consultants to identify the priority areas within a given period. Furthermore, the fact that the regional coordinators and theology consultants are themselves Africans helps to deflect the notion that this is a 'foreign' project. The RGGs have the autonomy to focus on the issues that they regard as pressing for their region. To a large extent, this overcomes the top-down approach.

One of the most notable achievements of EHAIA has been the transformation of theological thinking on the HIV epidemic in Africa (Chitando, 2008). Obviously, a lot of work remains to be done to ensure that churches move from denial and condemnation to effective responses. Through its workshops, conferences, seminars and publications, EHAIA has challenged churches in Africa to see the HIV epidemic in a new way. This has gone some way towards addressing the challenge of stigma and discrimination. Musa Dube, EHAIA's first theology consultant, edited the hugely popular, *Africa Praying: A Handbook on HIV/AIDS Sensitive Sermon Guidelines and Liturgy* (Dube, 2003). Many church leaders have utilized the book to encourage their congregations to support people living with HIV, and to be part of the solution.

Alongside other players in the ecumenical field, who include Churches United Against HIV/AIDS (CUAHA), Medical Assistance Programme (MAP) International, Lutherans, Anglicans, Catholics and others who are actively involved in HIV and AIDS work, EHAIA has transformed the churches' attitudes towards the epidemic. It has challenged churches to rethink their mission in the context of HIV. EHAIA has called for a paradigm shift that entails appreciating that HIV is at the heart of what it means to be churches at this moment in history. Thus:

> When applied to the crisis of AIDS, one could argue that it represents a *kairos* for the church. This is not to overlook the tragic and horrific dimensions of the disease. It means that the church ought to discover in the crisis engendered by the epidemic a radical understanding of its identity and commitment to its mission to renew humanity as well as facilitate the advent of a new community of faith.
>
> (Orobator, 2005, p. 123)

EHAIA has played a major role in mobilizing African church leaders to the struggle against the HIV epidemic. After having transformed the African churches' theological interpretations of the epidemic, EHAIA has challenged church leaders to be actively involved in prevention, treatment, care and support. Through conferences, workshops, seminars and publications, EHAIA has convinced many church leaders that they must play a major role in addressing the epidemic. As a result, many bishops and other church leaders have ensured that their churches are strategically placed to respond to HIV and AIDS. For example, in Zimbabwe, Bishop Xavier Chitanda, leader of the Union for the Development of Apostolic Churches in Zimbabwe (Africa) (UDACIZA), has sought to transform harmful cultural practices within Apostolic churches after having attended training sessions. UDACIZA is the acronym of the grouping of Apostolic churches that seek to respond effectively to the HIV epidemic.

EHAIA workshops, seminars and publications seek to raise awareness of the church leaders' responsibilities in the time of HIV. They are designed to cultivate interest in becoming agents of change. In addition, the workshops, conferences, seminars and publications empower church leaders to articulate the church's role in responding to the epidemic. The multiplier effect is felt when the trained church leaders in turn train others.

EHAIA has encouraged the formulation and application of HIV and AIDS policies by churches and church-related institutions. Such policies are important instruments for undermining stigma and discrimination. These policies are formulated at the local level and differ from one context to another. They are very effective in challenging negative attitudes towards people living with HIV. A good example is how Daystar University in Nairobi, Kenya, adopted a policy in 2007. Before implementing the policy the university demanded that prospective students and staff had to be HIV negative to be admitted or employed. In their 2007 policy they stated that they would no longer discriminate on the basis of a person's HIV status. In the specific case of church-related academic institutions, this has ensured that the tendency to require students and staff to be HIV negative in order to be admitted or hired has fallen away. This is a major success as it brings churches in line with human rights and legal requirements. EHAIA also encourages churches and church-related institutions to set up HIV and AIDS units, as well as putting aside human and financial resources for the work of these units. These funds are mobilized at the local level, although in some instances churches have been supported by external partners.

The transformation of the attitudes of religious leaders towards people living with HIV has seen a number of denominations apologizing for not having adopted a more proactive stance during the early phase of the epidemic. They have pledged to do more to ensure that their churches become safe and welcoming spaces to the infected and affected. Although stigma remains rife, with some continuing to regard the HIV epidemic as a sign of God's judgement, there has been a marked improvement in the churches' involvement. However, stigma remains a major challenge. Some religious leaders, especially within the Pentecostal sector, continue to associate HIV with personal sin. For example, in Harare, some Pentecostal preachers continue to preach that those living with HIV have questionable lifestyles (interview with Rev Maxwell Kapachawo, Harare, 13 February 2009).

Nyambura J. Njoroge, EHAIA Project Coordinator, observes that the HIV epidemic necessitates the emergence of a new style of leadership within the African churches. She argues that such leadership has to be inclusive and prophetic. It has to be sophisticated enough to be aware of the need to empower both the girl-child and the boy-child. For Njoroge:

> We need leadership that is willing to work with people, including children, youth and the very poor so that together we can become agents of change in our families, churches and communities. Such transformed and empowered leadership will be able to view injustices critically, reflect upon them and work towards their eradication

in the community. Transformed and empowered leadership will be prepared, above all else, to dismantle sexism, patriarchy, clericalism and hierarchy within the body of Christ.

(Njoroge, 2008, p. 192)

At a time when most religious leaders interpreted sin narrowly in terms of personal limitations, EHAIA drew attention to the structural sins that increase vulnerability to HIV. In particular, it highlighted how the epidemic thrives in contexts of poverty and gender inequality. It challenged Christians to actively oppose the factors that continue cycles of poverty and gender inequality. Churches in Africa have been encouraged to engage in critical social analysis to identify factors that increase vulnerability to HIV. As a result, a number of church leaders have spoken out on issues related to Africa's marginalization in the global economic order. The example of the former Archbishop of Cape Town, Njongonkulu Ndugane (2003), illustrates this point. He has prophetically challenged developed countries to contribute towards the response to HIV.

The HIV epidemic vividly illustrates the gap between the rich and the poor, and draws attention to those whose lives count, and those whose lives do not appear to matter. If we focus on the theological message, Christians should be called upon to side with those on the margins of society, following the example of Jesus who associated with the poorest of the poor. From a social point of view, this should be concretized into action and in their advocacy role, they must pressurize drug manufacturers not to put profit before human lives. At a time when antiretroviral drugs have become more affordable to the majority of people living in Europe and North America, many Africans continue to die unnecessarily. EHAIA has mobilized churches in Africa to call upon pharmaceutical companies to place absolute priority on human lives. According to UNAIDS (2008, p. 19):

> Despite the existence of affordable medications, too few people living with both HIV and tuberculosis are receiving treatment for both conditions – The failure to make optimal use of existing diagnostic and treatment regiments results in considerable illness and death.

The tragedy is that people continue to die at a time when HIV has become a manageable chronic disease in the Global North. Poverty has been the major reason why this state of affairs persists. Religious leaders therefore have a major role to play in insisting on the rights of the poor to receive life-prolonging drugs.

In addition, it is important to underline the fact that churches in Africa are beginning to make HIV an important governance issue. Although it has taken long for HIV and AIDS to become a major political issue in Africa (de Waal, 2006), it is clear that the quality of leadership available at the national level has a bearing on how particular countries respond to the epidemic. Denialism, centralization of resources, corruption and other factors tend to undermine effective responses to the epidemic. Building on the churches' contribution to the democratization of Africa in the 1990s (Gifford, 1995), EHAIA encourages churches to call on African governments to provide effective leadership in response to the HIV epidemic. While a longer narrative is required to do justice to EHAIA's efforts in Africa, it suffices to note that it has reached thousands of theologians and church leaders since its formation. It operates in all the linguistic regions of Africa (Anglophone, Francophone and Lusophone), as well as operating ecumenically. It is funded by different organizations, including those in the Global North and African churches. While it falls under the WCC structures in Geneva, it has an International Reference.

It is evident that AIDS has had an indelible impact on African politics. In many Southern African countries, AIDS has been responsible for the death of some politicians. Economic growth is stunted as productivity is reduced, there is greater demand on public health, democratic reversals might occur and potential voters could be affected by the epidemic. Weak governance as a result of HIV and AIDS might instigate instability (Chirambo, 2008, p. 30). Furthermore, it continues to place a strain on the resources of most governments who were already struggling to provide social services before the epidemic. Church members of the WCC are also promoting a campaign in support of moral change in many African countries. Churches in Africa must continue to pressurize their governments to reduce spending on luxury vehicles, building status symbols and engaging in wars at a time when their citizens need their health delivery systems to be delivered from intensive care. Through promoting such prophetic action, the WCC seeks to empower churches in Africa to occupy political space.

It must be conceded that the prophetic ministry of the church in relation to HIV has not been a spectacular success. Many church leaders do not possess the requisite political and economic literacy to challenge politicians. One major strategy that has shown a lot of promise has been to get church leaders to become members of the country-coordinating mechanisms (CCMs) of the National AIDS Councils (NACs). In a number of countries, church leaders have been appointed to head the NACs.

What is instructive to note here is that the church leaders act locally. Churches are being encouraged to make a difference at the local level, rather than wait for some transformation in the global economic order.

Given the notable presence of Christians in most parts of the continent, they clearly have a major role to play in influencing their governments to devote more resources to the struggle against the HIV epidemic. EHAIA has encouraged Christians to challenge their governments to respond quickly and effectively. In particular, it persuades church leaders to do justice to their prophetic role. It challenges church leaders not to allow local elites to abuse resources. The HIV epidemic has provided church leaders with an opportunity to repent, and to demand that ruling elites promote abundant life for all. Patrick Ryan has described the collusion between church leaders and indigenous elites in Africa:

> By and large the leadership of our church does not have a strong prophetic tradition. Church leaders have attended rallies of ruling parties; they have prayed sycophantically and celebrated domesticated liturgies that entrenched rather than challenged the wrongdoing of elites; they have accepted cash donations without enquiring into their source; they have not demanded transparent and accountable election procedures, fixed terms of office, and monitoring procedures for themselves; they have facilitated church structures that in some ways parallel the hegemonic and authoritarian civil structures; they have withheld offering institutional and legal protection to grassroots activists being persecuted for their justice ministry; they have remained silent when confronted with issue after issue that called for an urgent public response. Most disturbingly was the failure to discern and denounce the disproportionate suffering of women as victims of male power.
>
> (Ryan, 1998, p. 16)

In line with its quest for social justice, EHAIA places emphasis on the full participation and active involvement of people living with HIV in the churches and communities. Through the acknowledgement that 'the church is living with HIV' and that we inhabit 'a world living with HIV', EHAIA has sought to promote the visibility and appreciation of people living with HIV in churches. Its collaboration with the African Network of Religious Leaders living with or personally affected by HIV and AIDS (ANERELA) seeks to encourage African churches to welcome people living with HIV.

ANERELA national chapters work closely with EHAIA regional coordinators. There have been some notable achievements in terms of 'breaking the conspiracy of silence' (Messer, 2004). In Zimbabwe, Maxwell Kapachawo, the national coordinator of the Zimbabwe Network of Religious Leaders living with or personally affected by HIV and AIDS (ZINERELA) has become well known for his advocacy work. He has raised the community's awareness of the epidemic, and has contributed to the struggle against stigma and discrimination by declaring that he is a religious leader living with HIV. Some religious leaders in other countries have also courageously shared their testimonies. Although some Pentecostals continue to associate HIV with personal sin, there is greater awareness and acceptance of the reality of HIV across the different denominations and different religious groups.

The active participation of people living with HIV in church activities has implications that go beyond the religious sphere. As they have become aware of the support of their religious leaders, people living with HIV with church backgrounds have joined forces with other activists from civil society to demand their rights. It is envisaged that, as with the Treatment Action Campaign (TAC) in South Africa, the active involvement of people living with HIV will transform governance in Africa (http://www.tac.org.za/community/). EHAIA encourages church-based activists to join forces with other civil society actors to ensure that communities mount effective responses to the HIV epidemic.

EHAIA promotes the active involvement of people living with HIV because it is convinced that they are best placed to provide guidance in the development and implementation of HIV and AIDS policies. It goes therefore against the dominant paradigm where those with financial and political power make decisions on behalf of others. This 'preferential option for the infected and the affected' seeks to ensure that the experiences and opinion of those who are directly affected by the epidemic are taken seriously.

By seeking to implement the Meaningful Involvement of People living with HIV (MIPA), EHAIA endeavours to avoid the pitfalls of top-down development implementation. This is where outsiders purport to come with ready-made solutions to locals. Instead, MIPA allows those who are directly infected and affected to influence policy and practice. This strategy has seen greater emphasis being put on the acquisition of antiretroviral drugs, for example, where previously political leaders would have regarded the purchase of new vehicles as a priority.

It has become clear that the HIV epidemic has hit African women the hardest. Women's health and well-being have been compromised by the epidemic (Phiri & Nadar, 2006). Women are disproportionately affected by HIV. HIV infection is highest among young women. In addition, women are the primary caregivers for the sick and dying, accounting for two thirds of all caregivers. Women who are widowed as a result of HIV face social ostracism or destitution (UNAIDS, 2008, pp. 22–3). In military conflict, they are at risk of infection through rape. EHAIA encourages African churches to challenge patriarchy and its death-dealing practices. It calls upon them to side with women and children, and to work for conversion among men. This entails re-reading the Bible to promote the liberation of women, and to deconstruct aggressive masculinities. In the Plan of Action, African churches undertook to do the following in relation to gender:

1. We will challenge traditional gender roles and power relations within our churches and church institutions which have contributed to the disempowerment of women, and consequently to the spread of HIV/AIDS.
2. We will combat sexual violence, abuse and rape in homes, communities, schools and conflict/war situations.
3. We will address gender roles and relations in families that contribute to the vulnerability of women and girls to HIV infection.
4. We will support organizations that help young women to negotiate safer sexual relationships (Plan of Action, 2002, p. 11).

African churches and communities have not readily bought into the discourse on gender transformation in the wake of HIV. In particular, some male church leaders have regarded this as part of the larger 'Western' conspiracy to undermine African culture. Others have charged that changing gender relations goes against 'real Christianity' since there are biblical passages that support the submission of women. However, training sessions have led many church leaders to become advocates of gender justice.

Due to historical factors, African churches have tended to depend on 'mother' churches in Europe and North America for financial support (Chitando, 2007, pp. 34–5). EHAIA seeks to promote the utilization of local resources in the response to the HIV epidemic. Although there are sound historical, economic and theological reasons for overseas churches to support African churches, there is need to undo the dependency syndrome that has developed. Mobilizing local resources is a key strategy in providing effective responses to the epidemic. This has clear political implications as churches that are independent financially can interrogate oppressive actions by governments.

The WCC and HIV in Africa: reflections and conclusion

The foregoing sections have highlighted the achievements of the WCC in responding to the HIV epidemic in Africa. Despite the commendable progress that has made, there are a number of drawbacks in its efforts to ensure that communities of faith provide an effective response. This section draws attention to some of these factors.

As an ecumenical body, the WCC is well placed to mobilize its members to respond to the HIV epidemic. However, it does not have the capacity to ensure that its members do actually implement the programmes it initiates. It is entirely up to the churches on the ground to do so. While many churches have taken up the struggle against HIV, others have only paid lip service. The inability of the WCC to ensure compliance by its members results in limited success in the response to the epidemic. This critique is also applicable to the WCC contribution to political issues in Africa and other parts of the world (Mapuranga and Chitando, 2008).

The WCC has trained thousands of activists in different parts of Africa. These have become motivated workers in the struggle against the epidemic. However, there is limited follow-up. Trained individuals who are contributing to the struggle often feel isolated and even deserted. After the training workshops, they embark on activities to mobilize their communities while the WCC personnel who conducted the initial training moves on to other countries. This dimension compromises the WCC's interventions in Africa.

Like other ecumenical bodies, the WCC places too much emphasis on voluntarism. Whereas many non-governmental organizations involved in HIV and AIDS work pay per diems and other allowances to trainers and activists, the WCC continues to appeal to Christian charity. Individuals and organizations that collaborate with the WCC are expected to follow the charitable Christian message. This is a major weakness as many dedicated individuals end up pursuing other income-generating projects. As a result, WCC-inspired projects tend to have a limited life span.

Whereas the WCC had a very clear position regarding apartheid, for example, the changing global political order has left it struggling to articulate its position. Since the struggle against HIV is closely related to the struggle against poverty, inequality and discrimination, there is need for the WCC to sharpen its voice on these matters. However, it is envisaged that the WCC will rediscover its clear prophetic voice and provide the required leadership. This leads us to come to some conclusions.

The WCC's HIV intervention in Africa raises some key issues regarding the interface between religion and development. In particular, it illustrates how religious organizations occupy political space in Africa. Whereas churches in Africa initially regarded HIV as a non-theological issue, the entry of the WCC at the turn of the last millennium changed the situation. This external stimulus has empowered many churches to respond to HIV. As a global ecumenical body, the WCC possesses many advantages. It is able to bring together actors from diverse theological and ideological backgrounds and its status as a Geneva-based organization grants it political and economic capital. In addition, the WCC's previous engagement with the anti-apartheid struggle grants it some legitimacy, especially in Southern Africa. In order to situate the WCC's engagement with HIV it is therefore necessary to appreciate its previous response to institutional racism. This serves as a reminder that understanding the historical context is crucial if one is to grasp the interface between religion and development in any given situation.

It is also vital to understand the WCC's strategy of influencing theological transformation regarding the HIV epidemic in Africa. Recognizing that a top-down approach in development work is problematic, the WCC has sought to influence change through its member churches. To this end therefore, the WCC tried to develop a different approach and started to promote bottom-up techniques like local ownership of church-based HIV programmes. This has enabled the organization to meet many of its stated objectives.

One of the WCC's greatest advantages in its HIV interventions in Africa is that it eschews fundamentalist positions. Its openness enables it to collaborate with different organizations. Its workshops, conferences and seminars bring together actors of diverse ideological persuasions. In addition, the ecumenical platform eliminates denominational rivalry and reaches a wider audience. However, the WCC has not been able to totally eliminate the challenges associated with externally inspired development programmes/interventions. The major challenge is that the WCC acts as a catalyst and withdraws. In reality, fired-up activists do not always have the skills to negotiate the local context after the facilitators leave. This serves as a reminder to researchers within the field of religion and development to take cognisance of the local context.

Finally, the focus on the WCC's HIV intervention in Africa problematizes most of the popular terms in the discourse on religion and development. The WCC may at once be regarded as a development agency, religious entity, civil society group, local activist body and international ecumenical organization. Each one of these labels would be applicable depending on the context and mode of action that the WCC had undertaken at a particular historical moment.

References

Adogame, A. & Chitando, E. (2005), 'Moving Among Those Moved by the Spirit: Conducting Fieldwork within the New African Religious Diaspora' *Fieldwork in Religion*, 1(3), pp. 253–70.

Banana, C. S. (1996), *Politics of Repression and Resistance: Face to Face with Combat Theology*, Gweru: Mambo Press.

Beckmann, N. (2009), 'AIDS and the Power of God: Narratives of Decline and Coping Strategies in Zanzibar' in Becker, F. & Wenzel, P. G. (eds), *Aids and Religious Practice in Africa*, Leiden: Brill.

Chirambo, K. (ed.) (2008), *The Political Cost of AIDS in Africa: Evidence from Six Countries*, Pretoria: Idasa.

Chirimuuta, R. C. & Chirimuuta, R. J. (1987), *AIDS, Africa and Racism*, London: Richard Chirimuuta.

Chitando, E. (2007), *Living with Hope: African Churches and HIV/AIDS*, vol. 1, Geneva: World Council of Churches.

Chitando, E. (ed.) (2008), *Mainstreaming HIV and AIDS in Theological Programmes: Experiences and Explorations*, Geneva: World Council of Churches.

Cimperman, M. (2005), *When God's People Have HIV/AIDS: An Approach to Ethics*, Maryknoll, NY: Orbis Books.

Deneulin, S. & Bano, M. (2009), *Religion in Development: Rewriting the Secular Script*, London: Zed Books.

de Waal, A. (2006), *AIDS and Power: Why There is No Political Crisis – Yet*, London: Zed Books.

Dube, M. W. (ed.) (2003), *AfricaPraying: A Handbook on HIV/AIDS Sensitive Sermon Guidelines and Liturgy*, Geneva: World Council of Churches.

Esack, F. & Chiddy, S. (2009), 'Introduction' in Esack, F. & Chiddy, S. (eds), *Islam and AIDS: Between Scorn, Pity and Justice*, Oxford: Oneworld.

Gifford, P. (1995) (eds), *The Christian Churches and the Democratisation of Africa*, Leiden: E. J. Brill.

Iliffe, J. (2006), *The African AIDS Epidemic: A History*, Oxford: James Currey.

Lux, S. & Greenaway, K. (2006), *Scaling up Effective Partnerships: A Guide to Working with Faith-based Organizations in the Response to HIV and AIDS*, Geneva: Ecumenical Advocacy Alliance.

Mapuranga, T. & Chitando, E. (2008), 'The World Council of Churches and Politics in Zimbabwe: From the Programme to Combat Racism to Operation Murambatsvina (1969–2005)' in Kunter, K. & Schjørring, J. H. (eds), *Changing Relations between Churches in Europe and Africa: The Internationalization of Christianity and Politics in the 20th Century*, Wiesbaden: Harrassowitz Verlag.

Marshall, K. & Keough, L. (2004), *Mind, Heart and Soul in the Fight against Poverty*, Washington, D.C.: World Bank.

Messer, D. E. (2004), *Breaking the Conspiracy of Silence: Christian Churches and the Global AIDS Crisis*, Minneapolis: Fortress Press.

Ndugane, N. (2003), *A World with a Human Face: A Voice from Africa*, Geneva: World Council of Churches.

Njoroge, N. J. (2008), 'Daughters and Sons of Africa: Seeking Life-Giving and Empowering Leadership in the Age of HIV/AIDS Epidemic' in Hinga, T. M., Kubai, A. N., Mwaura, P. & Ayanga, H. (eds), *Women, Religion and HIV/AIDS in Africa: Responding to Ethical and Theological Challenges*, Pietermaritzburg: Cluster Publications.

Núñez, E. A. C. & Taylor, W. D. (1996), *Crisis and Hope in Latin America. An Evangelical Perspective*, Chicago: Moody Press.

Orobator, A. E. (2005), *From Crisis to Kairos: The Mission of the Church in the Time of HIV/AIDS, Refugees and Poverty*, Nairobi: Paulines Publications Africa.

Parry, S. (2008), *Beacons of Hope: HIV Competent Churches: A Framework for Action*, Geneva: World Council of Churches.

Paterson, G. (2005), *AIDS-Related Stigma: Thinking Outside the Box: The Theological Challenge*, Geneva: Ecumenical Advocacy Alliance.

Phiri, I. A. & Nadar, S. (eds) (2006), *African Women, Religion and Health: Essays in Honor of Mercy Amba Ewudziwa Oduyoye*, Maryknoll, NJ: Orbis Books.

Plan of Action (2002), *The Ecumenical Response to HIV/AIDS in Africa*, Geneva: World Council of Churches.

Ryan, C. (2003), 'AIDS and Responsibility: The Catholic Tradition' in Bate, S. C. (ed.), *Responsibility in a Time of AIDS: A Pastoral Response by Catholic Theologians and AIDS Activists in Southern Africa*, Johannesburg: St Augustine College of South Africa.

Ryan, P. (1998), 'The Shifting Contexts of Sin' in Ryan, P. (ed.), *Structures of Sin, Seeds of Liberation*, Tangaza Occasional Papers, No. 7, Nairobi: Paulines Publications Africa.

Schmid, B. (2007), 'Sexuality and Religion in the Time of AIDS' in Maticka-Tyndale, E., Tiemoko, R. & Makinwa-Adebusoye, P. (eds), *Human Sexuality in Africa: Beyond Reproduction*, Johannesburg: Fanele.

UNAIDS (2008), 'Report on the Global AIDS Epidemic: Executive Summary', Geneva: UNAIDS.

van Beek, H. (2006), *A Handbook of Churches and Councils: Profiles of Ecumenical Relationships*, Geneva: World Council of Churches.

Weinreich, S. & Benn, C. (2004), *AIDS – Meeting the Challenge: Data, Facts, Background*, Geneva: World Council of Churches.

World Council of Churches (WCC) (1997), *Facing AIDS: The Challenge, the Churches' Response: A WCC Study Document*, Geneva: World Council of Churches.

Conclusion

Reflections on Modernization without Secularization

Barbara Bompani & Maria Frahm-Arp

Reflecting on religion in the contemporary world forces us to rethink the failing paradigms of modernity and modernization theory, with their interpretations of secular development and the separation of religion from the public sphere and politics. If we think of the idea of modernity as a myth or a process, then we have to agree with Bruno Latour that 'we have never been modern' in the sense that this project has never been totally realized, and indeed perhaps it never can be (Latour, 1993, p. 1). Correspondingly, if we think in terms of the limits of modernization in non-Western contexts, then we have to consider that religion, in many places in the South, has never been anti-modern. On the contrary, starting with missions and the moments of first contact between the West and the rest of the world, religion has always been associated with modernity and its ideas and promises of progress and development. For example, in the kingdom of Buganda (today the nation of Uganda), the elite embraced Christianity as a vehicle to modernity, as one way of participating in the colonial culture and an opportunity to take advantage of its benefits (Mukasa & Gikandi, 1998). Similarly Ali Mazrui (1986) has shown the role that Islam played as a vehicle helping urban Africans move into elite and educated circles of power.

As this volume has demonstrated, the contraposition between modernity and religion, rooted in Western interpretations of secularism, does not ring true in non-Western societies and for this reason this relation needs to be rearticulated in theoretical and interpretative terms. Each chapter in this volume serves, in different ways, to highlight three key points that can aid us in better understanding the relationship between religion and what we think of as modernity in sub-Saharan Africa. Firstly, it emerges that religion holds different meanings in different contexts and that in non-Western contexts it is not a monolithic, rationally categorized entity but instead a 'messy' concept. Social scientists and development agencies need to take this into account in their analyses and actions. Secondly, religion has a specific and strong public role in Africa. Religions are exercising a powerful influence over the continent. Furthermore, religion is also a vehicle to give a public voice to local sentiments and to allow interaction with other forms of power, such as with politics. Thirdly, as above, rather than being perceived as an anti-modern agent in Africa, religion is often a vehicle of modernity per se.

Messy religion

According to Jonathan Crush (1995, p. 2) colonialism was all about 'gaining control of disorderly territory and setting loose the redemptive powers of development'. The normative power behind Western principles of development has leant towards the idea that development offers 'the power to transform old worlds, the power to imagine new ones' (Crush, 1995, p. 2). This is important in understanding the seemingly antipathetic relationship that exists between 'irrational' religion and Western perceptions of development. As both Ellis and ter Haar show in their chapters 'Development and Invisible Worlds' and 'The Mbuliuli Principle: What is in a Name?' we have to acknowledge a fear, in Western contexts or from a Western perspective, of the irrational; a reticence to engage in a constructive dialogue between the spiritual and the earthly. Although we can register a change in this trend and an emerging renewed interest in considering the role of religion in public life, we also need to acknowledge that this emergence is still quite limited. Until now, attempts to include religious organizations within development and political analyses has mainly revolved around the idea of 'normalizing', 'rationalizing' and 'categorizing' religious organizations in the logic of rational Western interpretations. As Crush states 'as an arena of study and practice, one of the basic impulses of those who write development is a desire to define, categorize and bring order to a heterogeneous and constantly multiplying field of meaning' (Crush, 1995, p. 2). As we have seen, there is much more to acknowledge and understand than simply the role of faith-based organizations in development, and mainline churches and religious leaders are not the only voices that can affect politics.[1]

Africa is a space in which the sacred, the transcendental and the spiritual are infused into every particle of life (ter Haar & Ellis, 2004; Gifford, 2008). Relationships, families, politics, economics, governments – all forms of power are understood to varying degrees in terms of spiritual powers. Studies on and into Africa need to search beyond the phenomenological issues at hand and place the theological ideas of the divine and the spiritual that inform the experience of the people they study into the context of their work. In doing so, they need to appreciate that the spiritual is further complicated by the facts that spirits can be both 'good' and 'bad', depending on one's point of view,[2] and that these are found in everything and everywhere. Just as religion can be an agent of development and positive change it can also be the force of destruction, violence and death in a community – leading to transformation and change. In a place of ongoing violence, as many parts of Africa are, and of disease, death is a reality that people have to cope with on a daily basis, something that van de Kamp explored in her chapter 'Burying Life: Pentecostal Religion and Development in Urban Mozambique'. In her chapter, images of death and dying offer members of the Universal Church of the Kingdom of God in Mozambique new, yet often problematic ways of coping with poverty in an urban setting.

This complicates how people in Africa work with religions and the various ways in which the spiritual is understood and practiced. Faith or religious affiliation, however it is perceived, does not operate in neat packages and what might at one time be a religious expression which has a positive impact on the lives of adherents may become a negative influence over time, or in other contexts. Religious ideas also slip out into all aspects of life, particularly in Africa where all is understood in terms of or in relation to spiritual power. Yet even this is changing and recent research into Pentecostal Charismatic Christian women in South Africa has shown that those who work in professional jobs have begun to separate their understanding of the world into spheres where spiritual powers are more dominant, for example in their personal relationships, and areas where spiritual forces have little or no power, for example through their work as chartered accountant dealing with financial statements (Frahm-Arp, 2009). Religions and religious organizations are nebulous and often difficult to define. Graveling for example examines this in her chapter 'Marshalling the Powers' when she talks about religions and their organizations and communities as being groups which:

> [do] not necessarily represent discrete bodies of people with coherent beliefs and secure boundaries. There may be enormous differences between the official policy of a church as set out by its head office and the activities of local branches; there may also be great divergence

between the doctrines and teachings of a local church and the beliefs, values and practices of its members. Attending or belonging to a church (or mosque) does not necessarily entail full acceptance of or compliance with its values and teachings; nor does it exclude acceptance of or engagement with other religious or non-religious traditions. (2010, pp. 198–9)

People also have multiple religious experiences as they move between churches or mosques, and between different forms of religion; for example young Muslims in Dar es Salaam may go to the mosque but also attend African Traditional Religious ceremonies as their families are still deeply involved in this form of religion (Janzen, 1994, p. 177). Various religious forms also appear 'messy' because different religions like Islam and Christianity have often been 'syncretized' and fused in various forms, or evolved into something culturally more acceptable, and in the process the symbols and forms of religious expression and even theologies have changed from those practiced in other parts of the world. The result of this has been the vast array of forms of Christianity, Islam and Hinduism which are practiced on the African continent and which differ, often quite markedly, from the forms of religious expression and theology found in Europe, the Middle East and Asia where the 'missionaries' of these religions originally came from. The Christianity practiced by African Independent Church members may therefore seem far removed from that of members of the Church of England (Bompani, 2008). This example expresses the complexity and flux found in just one religion and shows how we need to be careful of the categories and boundaries we create. Just as cultures are not static and discreet but are dynamic, fluid and flexible, so are religions.

Religions may be further complicated by the impact of and their interaction with globalization and urbanization. In Africa these factors seem to be at the heart of the contemporary popularity of Pentecostal Charismatic Christianity and the rise of Islam. But there has also been a re-emergence and re-shaping of what has been termed African Traditional Religion. One of the important things which all these religions do is help people to think about themselves and their world. In his chapter 'Muslim Shrines in Cape Town: Religion and Post-Apartheid Public Spheres'.

Tayob explained how the threatened destruction of old Islamic burial sites around Cape Town became a focal point that brought about a renewed sense of Public Islam and Islamic identity at the turn of the twenty-first century in South Africa. We hope that this volume has showed that both new and old religions become physical and symbolic spaces in which people are re-shaping themselves, their sense of self and their communities in a rapidly changing world, often in a fluid and dynamic way.

Religion and the public sphere

It is possible to think of two completely distinct and different ways of perceiving religion: as a personal and individual matter, or as a social and public component. Talking about religion, politics and development – particularly how they are shaped from below – means that we are engaging in a discussion about the public and the visible as well as the invisible world. Twentieth-century European thinking had largely regarded this world as private, but in other, non-Western, contexts this is regarded as just as public and powerful as any other visible power. In socio and political analyses the public space has been largely conceived of in terms of class, race, political groups and age, sometimes also gender. However, these categories do not always appear to be particularly helpful or even meaningful in understanding Africa societies, and there are other ways in which we could or should think about public space.

Usually religious powers or groups have largely been considered as relating to the public space in three ways. The first are movements that repudiate the public as millenarian movements do and they reject modernity. They are often driven by people who cannot survive in the modern context. The second are 'syncretic' movements which reach out to modernity and try to re-shape their own religious experiences and expression in a way that merges with the modern to create something new, often in a way that helps people cope with modern life. The third are those movements which seize the public space and try to control it by hijacking the power and responsibility of the state or military. The public space is one that also has various forces that control it – the community, the government and, in Africa, the military. Religions engage with these different public spaces, their actors and those who exercise power in them. Tied to the public space are economic forces, and religions engage with economic forces that shape networks and build economies which help or hinder people. People living in the countryside often experience religion, politics and development in different forms from that of many urban people. So location is important in understanding people's experiences and how people negotiate various forms of power. Access to resources is particularly important in understanding how people make decisions about the way they negotiate the difficulties in their lives and use religion, development agencies and political affiliation in strategic ways to improve their situation.

The public sphere is a central feature of modern societies, enabling people to move towards a common mind or cause, 'without the mediation of the political sphere, in a discourse of reasons outside power, which nevertheless is normative for power' (Taylor, 2007, pp. 190–91). As Taylor articulates it, the public sphere is the

> common space in which the members of a society are deemed to meet through a variety of media: print, electronic and also face-to-face encounters; to discuss matters of common interest, and thus to be able to form a common mind about these.
>
> (Taylor, 2007, p. 185)

When religion is strongly embedded in social structures, then it engages with 'rationality' and rational powers and it appropriates public spaces. As Cochrane explains in his chapter 'Health and the Uses of Religion: Recovering the Political Proper?' religion in developing countries is a reservoir of cultural autonomy, as moral authority and moral education, and as a major factor in the motivation of individuals. Paraphrasing Habermas, Cochrane presented religion as an idea of collective values.

> Despite its pathologies, religion is also generative in multiple ways, in part the reason for its force, persistence, and scale. Does one excise everything religious because of the pathologies, or is it worth paying attention to religion because of its generative capacities and strengths?
>
> (Cochrane, 2010, pp. 181–2)

Similarly Shulz in her chapter 'Remaking Society from Within: Extraversion and the Social Forms of Female Muslim Activism in Urban Mali' demonstrates how religion in Mali gives voice to women and generates a space for public debate. Female Muslims have become activists in Mali engaging in a particular form of 'politics from below' where they are not trying to generate or voice political protest, but to transform their personal and social reality.

In short, to understand religion in non-Western contexts we need to think in terms of *locus* and the non-separation between spaces and spheres. Oppositely to 'Western' and 'secular' perceptions of space, here there is no separation between spaces for rituals and worship and spaces for public and political debate. These spaces do not just overlap but fuse together to create one entity that is difficult to define. Similar spaces have also been observed in Latin America (Brusco, 1995). Understanding the meaning of religion in this context means losing the conceptual burdens of division or spheres. Taking people's systems of beliefs and their holistic approaches seriously is a real challenge for secular Western (and non-Western) researchers.

Rethinking religion and modernity

A critical approach to modernity is necessary for the understanding of 'who we are in the present' to recall Foucault's ideas (Foucault, 1984). It is difficult to define modernity as a concept, especially in the singular, because modernity is an increasingly recurrent and confused term of reference. Simply, in this book modernity has been considered as the images and institutions associated with Western-style progress and development in the contemporary world (Knauft, 2002, p. 18). Modernity has long been thought of as a teleological process, a movement towards a known end point that would be nothing less than Western-style industrialized modernity. These images and institutions do not need to be Western, but they remain at a certain level associated with Western social, economic and material development, in whatever way these are locally or nationally defined. Modernity in the contemporary developing world continues to be defined in association/comparison with the desire to improve social life by subordinating what is locally defined as backward and underdeveloped (Berman, 1994, p. 3). The desire to become modern is clearly not only and simply an intellectual academic speculation but is something tangible in the context of urban and rural spaces in Africa (Ferguson, 1999).

   To understand modernity we need to have a more flexible notion of what we mean by the term. Here Berman's concept of modernity as being a process in which change, flux, creativity, instability and disunity are key elements in the everyday fabric of life is particularly insightful (Berman, 1983). This means we can understand the spaces of emerging contradictions, multiplicity and shifting boundaries as actually being part of the process of modernity itself. Berman would argue that the very negation of the contemporary world which has been shown to be part of the African landscape, particularly with regard to corrupt governments, is also a part of the larger movement of societies and cultures towards becoming 'modern'. Central to this engagement with modernity is also a feeling of rootlessness, in that modernity itself cannot give people a sense of depth, meaning or a holding point around which to organize life. This book has also worked with the concept of modernity in these sorts of terms and highlighted what Berman failed to recognize, namely the importance of religion to people in the process of becoming modern. As many people find modernity an experience which destabilizes their world, so religions, particularly the contemporary forms of religion in Africa, have offered people modes of engagement through which to make meaning of their daily life experiences.

In this volume we did not follow postmodern interpretations. A prefix like 'post' implicates, per se, a concept of periodization, proceeding through ruptures or transformations, which modernity itself invented and delineated. The idea that the present is a point of transition or an element of a teleological transformation, at the edge of the new, located in relation to the historicity of events, and thus circumscribed by defined limits, belongs to a narrative of history (linear, progressive) that modernity itself inaugurated. From the 1960s post-structuralism has targeted the discourses that have authorized modernity as a project, interrogating the foundational concepts and narratives that underwrote it, like those of history as the linear and progressive unfolding of a *telos* (Foucault, 1984). Our final remarks are meant to highlight the relation of critique to the work of memory, for the danger today lies in forgetting that modernity was a Western 'invention'. There is an anthropological meaning of modernity as myth, 'which focuses on the story's social function: a myth in this sense is not just a mistaken account but a cosmological blueprint that lays down fundamental categories and meanings for the organization and interpretation of experience' (Ferguson, 1999, p. 14). A myth provides an understanding of the world, providing a set of premises that continue to shape people's experiences of their life. The myth of modernity has been characterized by the institution of a radically new form of sociality and subjectivity (Venn, 2000). Modernity was a myth formulated in a particular (Western) context in a specific historical moment. To understand the meaning of modernity in contemporary contexts (in our case to understand changes within African societies) it may be necessary to follow other paths.

Arjun Appadurai, for example, had suggested that it was necessary to rethink our understanding of modernity to take account of the many different sorts of *modern* cultural trajectories that researchers were documenting in different parts of the world (Appadurai, 1996). If non-Western cultures were not necessarily non-modern ones, then it would be necessary to develop a more pluralized understanding of modernity: not modernity in the singular but modernities in the plural, a variety of different ways of being modern. Would it be better to singularize or use the multiple form of modernity? This is undoubtedly a very appealing idea, but it immediately raises a number of problems. One problem is the meaning of the term 'modernity'. Once we give up the criterion of a singular modernity, then what does the term mean, analytically?[3] This book and its contributors suggest that it is not the meaning of modernity that has changed but the different actors and the ways in which they operate to reach their 'project of modernity' in different geographical and historical contexts. We would then use the word 'modernity' in the plural, maintaining the central meaning of the term,[4] but rather considering local specificities and exploring different local contributions to this idea. It is not the meaning of modernity that changes but the way modernity is pursued according to local interpretations. Thus there could be, for example, a South African modernity because there are South African actors pursuing modernity with local techniques, but the meaning of modernity in this context remains unaltered, and it still coincides with images and institutions associated with Western-style progress. Contextualizing local specificities means, to understand how to arrive to the *telos* with local resources, through local interpretations and within local realities. Modernity has previously failed to deliver on this and other promises; it is unable to reconcile the diversity of cultures. Societies and cultures cannot be understood as located along a continuum between 'pre-modern' tradition, on the one hand, and a Euro-centrically conceived modernity on the other.

It is necessary to consider modernity from a historical perspective to better understand the meaning of being or becoming modern. As part of this history, we have to consider how modernity has emerged in Western settings and how it connects socio-cultural transformation and orientations. These dynamics are mediated by cultural history and by the economic and political realities of what it means locally to be developed or experience (or desire) progress. Considering these trajectories and their relationship is crucial for an engaged and contemporary analysis. The idea of modernity offers in this way a mode of conceptualizing, narrating and experiencing rapid socio-political changes in the African state, bringing together evidence of the complexities and the ways that such processes have been understood by religious people and researchers. From the work in this book it is apparent that we need to consider modernity as a cultural space, a *locus* of social experience following a more recent literature (especially from an anthropological perspective) that claims that any theoretical interpretation

of modernity cannot be separated from culture and local interpretations. Seeing modernity as a cultural project allows us to understand the
ways local interpretations, of which religious understanding is a central
aspect, are used to render familiar the 'unfamiliar' paths to realize the
modernity project.

> Since modernity has not led to the wholesale convergence of societies
> and cultures, it is plain that there is nothing particularly 'natural' or
> inevitable about it. Modernity is not simply the logical outcome of an
> inevitable unfolding of structures and ideas. Rather, modernity turns
> out to have been cultural all along.
>
> (Moore & Sanders, 2001, p. 13)

Past interpretative models that did not consider local culture and its
importance failed to recognize the importance of cultural influence on
change. The adoption of these models creates a rigid divide between
Western and non-Western cultures and histories.[5] In this way development theory, as a part of the modernization project, constructs categories of persons and attributes greater emphasis to certain categories
and not to others on the basis of presumed effects that they already had
in other parts of the world.[6] There is a need for social science, politics
and development to abandon the conceptual interpretation of separate
spheres between religious and secular/social interpretations derived ultimately from the Enlightenment when outside of Western contexts; as
this volume has documented, this dichotomy loses coherence in the
context of African societies. As Martin asserted 'only by descending to
the quotidian and the empirical that one can observe the ways in which
such [religious] movements operate to empower individuals in new ways
and open up to them freedoms' (1998, p. 113). We need something of a
revolution of thought for the formulation of new theoretical approaches
and new methodologies that take into account, without misgivings, the
potentiality of religion in contemporary societies.

Notes

1. Or as Chitando's chapter showed, sometimes the top-down, rational approach
   of these organizations does not bring to the expected results. Chitando, E.
   'Sacred Struggle: The World Council of Churches and the HIV Epidemic in
   Africa'.
2. And also social and political results of the involvement of religious organizations have a variety of outcomes, both positive and negative for societies. This
   is the point expressed in different ways in Mallya's and Skinner's chapters.

Mallya, E. 'FBOs, the State and Politics in Tanzania'; Skinner, D. '*Da-wah* in West Africa: Muslim *Jama'at* and Non-Governmental Organisations in the Gambia, Ghana and Sierra Leone, 1960–90.

3. For support for this idea of modernity in the singular, see for example Ferguson, J. (2006), *Global Shadows*.

4. Modernity as the images and institutions associated with Western-style progress and development in the contemporary world.

5. That is the case followed, for example, by Anthony Giddens in Giddens, A. (1990), *Consequences of Modernity*.

6. See Knauft, B. M. (ed.) (2002), *Critically Modern*.

References

Appadurai, A. (1996), *Modernity at Large, Cultural Dimensions of Globalization*, Minneapolis: University of Minnesota Press.

Berman, M. (1983), *All that is Solid Melts into Air: The Experience of Modernity*, London: Verso.

Berman, A. (1994), *Preface to Modernism*, Urbana: University of Illinois Press.

Bompani, B. (2008), 'African Independent Churches in Post-Apartheid South Africa: New Political Interpretations' *Journal of Southern African Studies*, September, 34 (3), pp. 665–77.

Brusco, E. (1995), *The Reformation of Machismo: Evangelical Conversion and Gender in Colombia*, Austin: University of Texas.

Crush, J. (ed.) (1995), *Power of Development*, London: Routledge.

Ferguson, J. (1999), *Expectations of Modernity. Myths and Meaning of Urban Life on the Zambian Copperbelt*, Berkeley: University of California Press.

Ferguson, J. (2006), *Global Shadows. Africa in the Neoliberal World Order*, Durham, N.C.: Duke University Press.

Foucault, M. (1984), *The Foucault Reader*, New York: Pantheon.

Frahm-Arp, M. (2009), *Professional Women in South African Pentecostal Charismatic Churches*, Leiden: Brill.

Giddens, A. (1990), *Consequences of Modernity*, Stanford: Stanford University Press.

Gifford, P. (2008), *African Christianity and Its Public Role*, London: Hurst and Company.

Janzen, J. (1994), 'Drums of Affliction: Real Phenomenon or Scholarly Chimera?' in Blakely, T., van Beek, W. & Thompson, D. (eds), *Religion in Africa: Experience and Expression*, London: James Currey.

Knauft, B. M. (2002) (eds), *Critically Modern. Alternatives, Alterities, Anthropologies*, Bloomington and Indianapolis: Indiana University Press.

Latour, B. (1993), *We Have Never Been Modern*, Cambridge, Massachusetts: Harvard University Press.

Martin, B. (1998), 'From Pre-to Postmodernity in Latin America; The Case of Pentecostalism' in Heelas, P. (ed.), *Religion, Modernity and Post-Modernity*, Oxford: Blackwell.

Mazrui, A. (1986), *The Africans: A Triple Heritage*, London: Little, Brown.

Moore, H. L. & Sanders, T. (eds) (2001), *Magical Interpretations, Material Realities. Modernity, Witchcraft and the Occult in Postcolonial Africa*, London: Routledge.

Mukasa, M. & Gikandi, S. (1998), *Uganda's Katikiro in England*, Manchester: Manchester University Press.

Taylor, C. (2007), *A Secular Age*, Boston: Harvard University Press.

ter Haar, G. & Ellis, S. (2004), *The Worlds of Power: Religious Thought and Political Practise in Africa*, Oxford: Oxford University Press.

Venn, C. (2000), *Occidentalism. Modernity and Subjectivity*, London: Sage Publications.

Index

Africa
  African churches, 221–3, 228–35
  African development, 5, 7, 9, 25–9, 35, 45, 62
  African philosophy, 8, 26
  African politics, 6–7, 10, 13, 28, 30–1, 37–8, 46, 57, 76–7, 89, 110, 118, 120, 137–9, 221, 232–3, 240, 248
  African religion, 25–6, 30–2, 35–7, 46–7, 57, 78, 85, 119, 157, 199–200, 243
aid, 13, 48, 52–3, 104, 107–11, 116, 118, 120, 125, 146
  aid agencies, 47, 78, 84, 116, 141
  aid workers, 47
AIDS, 16, 145, 148–9, 182, 184, 218–25, 227–9, 232–6
ancestors, 37, 154, 156–7, 202–3, 213
Angola, 26, 226
anti-apartheid, 226, 237
apartheid, 11, 56, 61, 65–6, 226, 236
Appadurai, A., 247
Arab, 26, 75–6, 78, 88, 100, 110–11, 125, 132–3, 135
Asia, 2, 24, 66, 226, 243
autonomy, 188, 225, 228, 245

Bayart, J. F., 3, 7, 16, 57, 76, 78
belief, 3–4, 8, 27, 31–4, 38, 41, 49, 54, 134, 139, 180, 185, 190, 198, 202–3, 211, 213, 242–3, 246
  belief system, 6, 200, 211
Bible, The, 24, 31, 183, 202, 208, 213, 235
biomedicine, 187, 203, 211
bodily integrity, 187
body, bodies, 61, 155, 186–7, 212, 221
bureaucracy, bureaucratic, 5, 24–5, 29, 36–7, 60, 134, 175
bureaucratization, 9

capital
  economic/financial, 4, 11, 28, 32–3, 37, 49, 138, 163, 175, 237
  social, 44–5, 69, 71
  spiritual, 44–5
capitalism, capitalist, 9, 14, 23, 27, 29, 33, 44, 152, 153–4, 158, 204
  as magic, 153
Catholic, Catholicism, 2, 4, 10, 26, 45, 51, 112, 120–1, 139–40, 165, 201, 205, 207, 222, 229
Chabal, P., 9
Christianity, 7, 15, 23–4, 26–7, 36, 40, 49, 66, 68, 76, 78, 101–2, 107, 110, 119, 121, 127, 132–3, 135, 145, 152, 182, 185, 191, 199–200, 202–3, 209, 213, 219, 220–7, 231, 233, 235–6, 240, 243
  Charismatic Christianity, 2, 13
  Christian missionaries, 27
  Christian theology, 24
citizenship, 57, 175, 178
civilization, civilizing, 27, 34, 48, 80, 85, 90–1, 198
civil society, 6–7, 41, 76, 85–6, 131–2, 135–6, 138, 146, 175, 189, 191–2, 225, 234, 237
  civil society organizations, 131–2, 135–6, 137–9
civil war, 113, 127, 157, 159
cold war, 28
colonial, colonialism, 27–8, 30–1, 35–6, 42, 46, 51, 56–7, 62, 79–80, 82, 85, 90–2, 100–1, 106, 119, 121, 133, 135–6, 158, 161, 168, 198, 240–1
Comaroff, J., 5, 153, 166, 198, 204
Comaroff, J. L., 5, 153, 166, 198, 204
communication, 3, 12, 14, 30–1, 37, 47, 124, 176, 179–80, 183–6, 190, 197, 204, 213